From Cells to Chaos: Unveiling the Physics Behind the Heartbeat

Ravi

Contents

Chapter 1

Introduction

1.1 Motivation

The heart is one of the body's most central organs for life support, as it is responsible for circulation of blood through the whole body, which in turn transports nutrients and metabolic waste products between the individual organs. Cardiac muscle tissue forming blood-filled cavities contract rhythmically, and valves at the entrance and exit of the main chambers direct the stream of blood, altogether pumping blood from the atria to the ventricles and eventually through the entire body. Due to its vital role, the heart is very robust and adaptive to many demanding circumstances. Nonetheless, a disease that limits its pumping function is immediately life-threatening, mostly due to the dependence of the brain on a steady supply with oxygen. In addition to the heart's special anatomy, complex electro-chemical processes occur inside the muscle tissue on the molecular level. Each cell can be stimulated from its neighboring cells to elicit a so-called action potential (AP), a temporary excursion of the electrical trans-membrane voltage from its resting state, that in turn triggers surrounding cells. Macroscopically, the dynamics of these processes can effectively be described with the theory of excitable media, as wave-like excitation patterns spread through the heart, that provoke mechanical contraction of the individual muscle cells. Normal heart beat is characterized by regular and ordered contraction, the frequency of which is adapted to the current physical activity. However, in certain situations the heart can be driven out of this stable state of regular beating. Arrhythmias such as tachycardia and fibrillation show activation patterns with faster or even chaotic dynamics, respectively. During these phases the pumping efficiency is reduced dramatically, and the transition time for spontaneous return into sinus rhythm is often very long, leading to the brain taking damage, or in the worst case death.

From the theoretical perspective of physics and mathematics, the heart is a complex dynamical system, that can exhibit both, highly periodic and highly chaotic behavior. A fundamental concept of chaos theory is that the future evolution of a dynamical system can depend critically on small differences in the initial conditions. In the heart, this holds true for the dynamics of a single cell, where a small raise of the membrane potential over the excitation threshold elicits the generation of an action potential, as well as for the whole organ, where an unfortunately timed punch to the chest can cause a disturbance during a vulnerable phase of the heart beat and lead to life-threatening ventricular fibrillation (VF). On the upside, the same principle opens opportunities for novel control strategies of cardiac dynamics by means of small, appropriately timed and localized disturbances.

At the center of the complex and potentially chaotic dynamics of the heart lies the spread of excitation waves inside a substrate of muscle tissue, forming spatio-temporal patterns that are influenced by the 3D geometry of the heart as well as the dynamic patterns themselves. Many aspects of these processes are still poorly understood. Before novel concepts for control of arrhythmia can be successfully developed and brought into therapeutical application, it is key to further study and understand the evolution of regular and chaotic excitation waves, their interaction with each other and with the underlying inhomogeneous tissue of healthy and diseased hearts. Here, the challenge is to measure the dynamics with high spatio-temporal detail while it happens inside the living heart. The aim of this book is the advancement of the experimental tools to do so.

1.2 The mammalian heart

1.2.1 Anatomy and function

The mammalian heart consists of four chambers (fig. 1.1a). Right atrium (RA) and right ventricle (RV) pump blood coming from the body to the lungs, where it becomes enriched with oxygen, and is then pumped through the whole body by the left atrium (LA) and left ventricle (LV). The cardiac cycle consists of two phases. During diastole, the heart muscle is relaxed and the chambers fill with blood. The contraction phase is called systole. In normal heart beat, the so-called sinus rhythm (SR), a regular electrical excitation wave travels through the heart tissue. It originates from special pacemaker cells located in the RA, called the sinoatrial (SA) node, from where the excitation first spreads through both atria and stimulates them to contract simultaneously and further push the blood into the ventricles. At the heart's base, atria and ventricles are electrically isolated, the only conducting connection being the atrioventricular (AV) node located in between. It acts as a delay element for the excitation wave. This important function ensures that the ventricles do not contract before the atria have finished ejecting their blood. From the AV node, a branching network of fast conducting cells (Purkinje fibers) spread the impulses throughout the thick ventricular muscle tissue to create synchronized contractions. The pressure of the stimulated ventricular contraction is so high, that the valves at the entrances of RV and LV are forced shut and exit valves open, which directs the stream of blood and prevents back-flow. The AV node serves as a low-pass filter, in case the atria are in a state of quickened irregular activity, and also as a slow backup pacemaker, in the event of the SA node being inoperative.

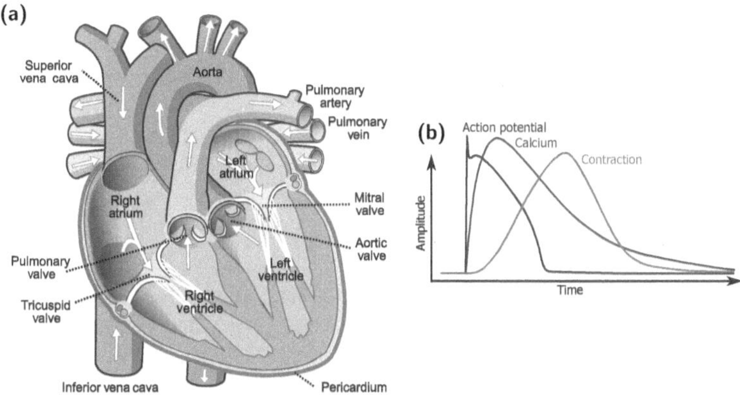

Figure 1.1 – Anatomy and dynamics of the mammalian heart. (a) Diagram of the human heart as an example of four-chambered mammalian hearts. Arrows indicate blood flow. Drawing by Eric Pierce.[1] (b) Time course of an action potential, Ca^{2+} transient and contraction measured in a rabbit ventricular myocyte at 37 °C. Schematic adapted from [31].

1.2.2 Electrophysiology of heart muscle cells

Like most cells, heart muscle cells (cardiomyocytes) actively maintain a specific intra-cellular mix of Na^+, K^+, Cl^-, and Ca^{2+} ions, whose concentrations largely differ from those in the extra-cellular interstitial fluid. Various specialized proteins (ion pumps, ion channels, ion exchangers) embedded in the cell membrane passively or actively regulate the desired ionic concentrations. These ionic concentration differences in sum constitute a voltage across the cell membrane, called membrane potential, which has a resting value of about -90 mV, as measured against the extra-cellular medium. Cardiomyocytes are excitable cells (like neurons, muscle cells, and some other cell types), meaning that the cell can be stimulated to generate an action potential, a rapid depolarization and following slow repolarization of the membrane potential (fig. 1.1b). The generation of the action potential is caused by a complex interplay of the various ion channels: When the membrane potential is raised above a certain threshold, voltage-gated sodium channels open, which allow even more sodium ions to flow into the cell, giving rise to the fast AP upstroke. Inactivation of the sodium channels and opening of delayed rectifier potassium channels results in a plateau phase due to roughly balanced flux of charges into and out of the cell. The influx of Ca^{2+} triggers release of more calcium ions from the sarcoplasmic reticulum within the cell — a process called calcium induced calcium release. In turn, this increase of intra-cellular Ca^{2+} finally causes the myofibrils to contract. The eventual repolarization to resting potential is achieved by the interplay of calcium,

---

[1] Drawing created by Eric Pierce and used unchanged under license CC BY-SA 3.0 [115], original file: `https://commons.wikimedia.org/wiki/File:Diagram_of_the_human_heart_(cropped).svg`

potassium, and chloride channels. During the plateau phase of the action potential, the cell is absolute refractory, i.e. it cannot be excited to produce another action potential. During repolarization phase, the cell becomes more and more excitable to larger-than-usual stimuli. At the end of an action potential the membrane potential returns to the initial resting value. However, the absolute ion concentrations have changed during the AP. Normal levels are actively restored during diastole by the ion pumps and ion exchangers.

1.2.3 Tissue dynamics

In the heart, muscle cells are arranged in fibers and their cytoplasms are connected by gap junctions, which allow the action potential to stimulate neighboring cells, resulting in excitation waves traveling through the tissue. The muscle fibers form sheets, within which the conduction velocity (CV) is largest in the direction parallel to the muscle fibers. The sheets are only loosely connected, therefore the CV is smallest in the sheet-normal direction.

During regular sinus rhythm, the excitation waves follow the pathways laid out by the anatomy. Most importantly, excitation passing through the AV node is spread in the ventricles around the apex by the electrical conduction system. This ensures that the uniting wave front freely travels towards the base, where it vanishes and gives way for the excitation wave of the following heart beat. If this process is disturbed, all kinds of irregular self-sustained recurring activation patterns can arise, that can dominate and suppress the typically slower sinus rhythm. Excitation waves may follow so-called anatomical reentries, e.g. loops around big blood vessels or non-conducting scar tissue. Next to such structural properties, the spatio-temporal propagation is also influenced by dynamic processes. The relative refractoriness of the action potential gives rise to the phenomenon of slowing-down of an excitation wave, when it runs into the tail of a preceding wave. Other forms of interaction are redirection, blockage, and wave front breakup. A spiral wave that rotates around a self-created obstacle is a form of a so-called functional reentry. Fibrillation is a dynamic state, where many small spiral wavelets continuously run into the paths of each other, leading to annihilation and generation of new spiral waves.

As an example of current research in our Research Group Biomedical Physics (BMP), a proposed treatment technique termed LEAP (low energy anti-fibrillation pacing) aims to provide a more gentle and energy-saving option with less side effects, compared to the conventional defibrillation method [63]. The idea behind this novel method is that several small electrical shocks applied at the right time can successively drive the system back into stable sinus rhythm. Contrary, conventional defibrillation uses a single high-current shock, which stimulates each excitable cell, in order to force all into absolute refractoriness and eventually bring them back to resting state. Hereby, the chaotic excitation patterns that cannot terminate on their own are erased with high success rates, without the need to know the exact pattern of activity. In turn, the negative side effects of the harmful electrical current have to be accepted.

While the molecular processes underlying heart dynamics can be studied in detail on individual cardiomyocytes, and 2D cell-cultures of heart tissue allow to recreate some aspects of irregular and chaotic dynamics, the complex interplay of the three-dimensional spatio-temporal excitation patterns is fundamentally different on the original anatomy of the heart and can only be observed realistically in the whole organ. Numerical simulations are not yet good enough to reproduce all aspects in fine detail on a 3D geometry and in reasonable time. Furthermore, numerical findings need to be validated in the experiment.

1.3 How to study a whole living heart?

The most realistic platform to study living hearts is provided by so-called *in vivo* experiments, i.e. in the living animal. The main problem is that the heart is difficult to access and to observe inside the body. Open-chest experiments only grant view from one side and the red haemoglobin may interfere with optical measurement techniques. Creating and maintaining artificial controlled experimental conditions can also be difficult in the living animal.

To overcome these issues, two techniques have proven to be very useful. They have greatly helped to advance the understanding of cardiac dynamics. Firstly, *Langendorff perfusion* bridges the gap between in vivo and cell-culture experiments, as it allows to keep an isolated heart alive *ex vivo*, i.e. outside of the body [1, 47, 59]. Thus, it can be easily accessed and studied in a controlled environment for a prolonged time. Secondly, *optical mapping* enables recording of the spatio-temporal dynamics of the internal excitation waves on the epicardial surface in high resolution with cameras and fluorescent dyes [32, 67]. Using multi-wavelength fluorometry, optical mapping facilitates acquisition of multiple physiological variables at once, e.g. transmembrane voltage and intracellular calcium concentration [22, 25, 36]. These two methods, Langendorff perfusion and optical mapping, have become a kind of gold standard in cardiovascular and pharmacological research and will be introduced in more detail in the next chapter.

A major difficulty with optical mapping is its susceptibility to motion artifacts. To date, a workaround often used is complete suppression of the muscle contraction with the help of chemical substances. Hereby, however, examination of an essential characteristic of the heart—namely its pumping function—is rendered impossible. Another complexity comes with the 3D shape of the heart. In order to map the whole surface, multiple cameras are necessary. Quantitative analysis of the evolution of excitation wave patterns across a curved surface on the basis of individual 2D projections is limited, without additional precise knowledge of the 3D geometry and imaging properties.

In the past, several developments have successfully overcome the individual problems. Camera calibration and 3D surface reconstruction methods enable multi-camera 3D panoramic optical mapping of the entire *static* heart surface [7, 15, 24, 35, 42, 46, 51]. Motion artifacts can be compensated using ratiometric imaging techniques, which require extended hardware imaging capacities for acquisition of two wavelengths per view [11, 27, 58]. Alternatively, motion artifacts can be reduced in software by co-moving image analysis using numerical 2D motion tracking techniques [82]. Lastly, methods for 3D motion tracking facilitate optical mapping of a *contracting* heart surface, some of which require additional active hardware such as structured light projectors [68, 77], while others can perform passive tracking of *artificial* surface features (markers) in 3D with image-based stereo vision techniques [60, 87].

However, the application of these methods is still very demanding. Especially the methods requiring additional hardware are not easily extendable to full 360° panoramic mapping of the beating heart. Furthermore, the principle limitation of optical mapping to the heart surface has not yet been overcome. Dynamic excitation wave patterns inside the thick cardiac tissue, as well as the intramural structure and motion remain inaccessible with these optical methods.

1.4 Aims of this book

Overarching goals. To give a more comprehensive and detailed view of the spatio-temporal interplay between the different processes in the heart, a novel experimental Langendorff setup shall be developed, that facilitates multimodal mapping of the structural and dynamical properties of a contracting heart at high spatial and temporal resolution, on its surface and within. To this extent, suitable aforementioned optical methods shall be advanced, and accompanied with other non-invasive methods for *simultaneous* acquisition of internal structure, motion, and electrical dynamics.

Detailed aims. A Langendorff perfusion setup for pig and rabbit hearts, suitable for static 3D panoramic optical mapping of membrane voltage with four high-speed cameras already exists in our group, and serves as basis for the following developments:

- Development of a novel, purely image-based numerical 3D surface deformation tracking technique of the contracting heart, utilizing *natural* surface features.
- Integration of a medical 4D ultrasound device for simultaneous, live volumetric recording of intramural structure and motion.
- Integration of a commercial 256-channel electrocardiogram (ECG) device for simultaneous measurement of the internal electrical dynamics.
- The acquisition of all modalities shall be spatiallaly calibrated with respect to a common coordinate system.
- Acquisition of the static heart anatomy at high spatial resolution after the experiment using a CT (computed tomography) scanning device.
- Facilitation of easy future extension with additional cameras for full 360° mapping, multi-wavelength measurement of intracellular Ca^{2+} concentration, and (if necessary) ratiometry, using the respective established techniques.

Challenges and considerations. While the ultimate aim is 360° optical mapping of a beating heart, this is not possible with just four cameras. The techniques to be developed shall not pose a principal limitation on the number of cameras in future applications. The main challenge is the integration of the additional equipment for ultrasound and ECG acquisition into the perfusion bath, without obstruction of the optical mapping. The bath further limits the positioning of the cameras. Dual-wavelength fluorometry of intracellular calcium is also desirable for the future, but this technique is already established and does not need further development from my point of view. This work therefore focuses on the development and integration of the other techniques. The same holds true for ratiometric imaging. If necessary, it can be added at a later stage. Our hope is, that the reduction of motion artifacts by motion tracking proves to be good enough to render extra hardware needed for ratiometry dispensable.

1.5 book structure and methods overview

The fields of work of this book span a very diverse set of topics. Therefore, the document is structured into mostly self-contained chapters, compiled in two parts. Each chapter has its own introduction, methods, and discussion section. Part I deals with the build of the experimental system and accompanying computational methods for high-resolution 3D panoramic optical mapping of static, motion-inhibited whole hearts. This is the basis for part II, which extends the experimental system with means for simultaneous recording of the additional modalities.

The four chapters in part I are ordered such that they build upon each other. Chapter 2 introduces the experimental techniques of Langendorff perfusion and optical mapping, and discusses the challenges of motion artifacts in more detail. Following, the previously existing ex vivo perfusion setup with multi-camera optical mapping of pig and rabbit hearts is presented. Various necessary improvements will be shown and discussed. Chapter 3 covers the topic of calibrating the imaging characteristics of all used cameras and lenses, and the validation thereof in the context of our multi-camera setup. Building upon this, chapter 4 details the methods used for 3D reconstruction of the heart surface from multiple silhouette images. This multi-step processing toolchain has evolved during my work on this book and before. It is validated by application to a rigid phantom of known size. I also present an alternative approach based upon a technique called Structure from Motion. Chapter 5 has the topic 3D panoramic optical mapping of static hearts. Individual images of the multiple calibrated cameras are projected and combined into polar maps of the whole epicardium. This allows to study the dynamical processes visible at the surface in high detail by analysis of action potentials, activation maps, phase singularities, and apparent conduction velocity on the curved surface.

The chapters of part II can be read in arbitrary order. While each modal extensions does not depend on the others and can be used individually, they are designed for non-interfering simultaneous application. Researchers can choose to use only one extension or all at once, depending on their needs. Chapter 6 presents methods for high-resolution CT-scanning the anatomy of the heart after the experiment. Chapter 7 extends the experiment with a novel, purely image-based, and marker-free 3D motion tracking method for the contracting heart surface. It bases on a combination of the static 3D geometry reconstruction with established 2D motion tracking of natural features. Additionally, a technique for estimation of the excitation light field is developed, that enables a reduction of motion artifacts, similar to ratiometric methods, but using just one wavelength. Chapter 8 covers integration of a 4D ultrasound system for simultaneous acquisition of volumetric videos. A new bath is constructed with an acoustic window in the bottom. For spatial alignment with the optical reconstructed 3D surface mesh, the probe's pose needs to be calibrated. An in situ calibration method is developed that does not require external tracking. Chapter 9 deals with simultaneous acquisition of multi-channel ECGs, in order to facilitate development and validation of new methods for non-invasive mapping of the electrical activity in the heart.

Chapter 10 concludes this work, reviews the result as a whole and gives an outlook for future application and development.

Part I

High-resolution 3D panoramic optical mapping

Chapter 2

Ex vivo optical mapping of isolated hearts

2.1 Introduction

In this chapter the methodologies and equipment of an experimental setup for ex vivo optical mapping of isolated hearts are introduced, which form the basis for this book. At first, the two main methods on which the experiment is based are intro-duced, namely *Langendorff perfusion* and *optical mapping*. Langendorff perfusion is a technique for keeping whole hearts of mammals (like pig, rabbit, and mouse) alive *ex vivo*, i.e. isolated from the body of the animal, therefore enabling exper-iments with direct access to the organ. The so-called optical mapping technique uses fluorescent dyes and cameras for contactless measurement of the heart tissue's key dynamical properties, like electrical trans-membrane voltage and intra-cellular calcium concentration. The challenges posed by moving samples to optical mapping are briefly introduced and discussed. Lastly, following a description of the existing ex vivo optical mapping setup available in our group at the beginning of my doctor-ate work, various necessary improvements to the experimental setup are presented, that were developed for the purpose of this book.

2.1.1 Langendorff perfusion

Langendorff perfusion makes it possible to keep a whole intact, isolated mammalian heart alive. This technique was developed by Oscar Langendorff and originally pub-lished in 1895 [1]. Since then it has become a standard method in cardiac research. The heart is perfused with blood or an physiological replacement solution, which supplies the organ with the necessary nutrients, electrolytes and oxygen, in order to stay alive. Using this technique, the heart can be studied *ex vivo*, i.e. outside of its 'natural environment' (the animal's body), but still under reasonably phys-iological conditions, and easily accessible. Langendorff perfusion therefore bridges the gap between *in vivo* animal experiments on the one hand, and *in vitro* cell culture experiments on the other hand. In vivo, the heart can be studied in its

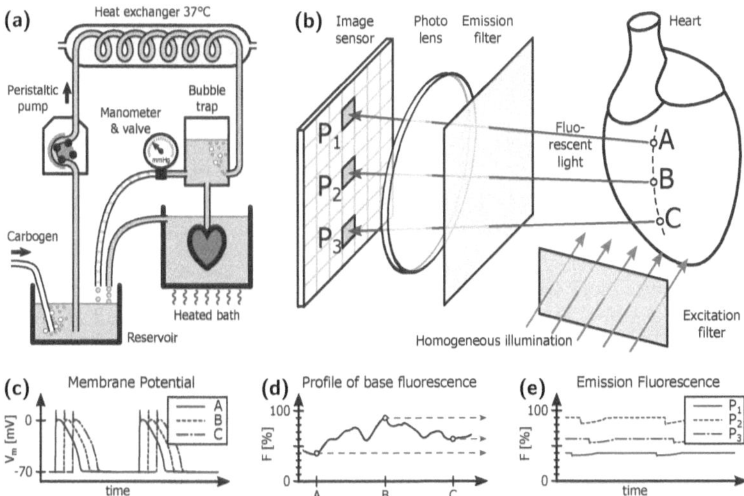

Figure 2.1 – Langendorff perfusion and optical mapping principles. (a) Schematic of a constant-pressure Langendorff perfusion setup. (b) Schematic of optical mapping principle on heart tissue loaded with potentiometric fluorescent dye. (c) Electrical membrane potential of tissue points A, B, C. (d) Spatial variation of base fluorescence in tissue (profile along dashed line in panel b). (e) Optical timeseries recorded with image sensor, show small fractional change of emission fluorescence. **Remark:** panels c to e show illustrative artificial data.

natural physiological environment, but cannot be accessed directly, except with invasive open chest surgery. Cell cultures of cardiomyocytes provide direct access in a controlled environment, but in vitro conditions are still quite unnatural, making it difficult to translate findings to the organ level. Furthermore, cell cultures only permit to study the dynamic excitation wave phenomena in two dimensions.

The main objective of the Langendorff perfusion method is to establish a circulation of the heart muscle with the perfusion solution, independent of the hearts own pumping action. Therefore, the perfusate is pushed through the heart's vasculature using an external pump or the gravitational pressure of a high-hanging reservoir. The heart is supplied in a retrograde fashion, meaning that the perfusion system is connected to the aorta. The aortic valve is forced shut by the pressure, directing the perfusion solution into the coronary arteries, which originate from the root of the aorta. Hereby proper circulation of the whole heart muscle is ensured, making it possible to perform experiments also in cases where the heart's own contractions are not present or suppressed.

A typical Langendorff setup with constant-pressure perfusion of the heart is shown in fig. 2.1a. It consists of a reservoir, gas bubbling lance, pump, heating, bubble trap, pressure regulator, and optional bath. The perfusion solution should be roughly isotonic with interstitial fluid, hence it typically contains sodium (Na^+), potassium (K^+), chloride (Cl^-), calcium (Ca^{2+}), magnesium (Mg^{2+}), hydrogencarbonate (HCO_3), phosphate (PO_4), as well as glucose ($C_6H_{12}O_6$) in physiological concentrations. The electrolytes play an important role in the generation of the cells' electrical action potentials, while the glucose serves as energy source. Hydrogencarbonate and phosphate are used as a buffer. The perfusate is bubbled with Carbogen gas, a mixture of 95% O_2 and 5% CO_2. While the oxygen obviously is needed for energy production of the muscle cells, the carbon dioxide regulates the solutions pH value. A bubble trap prevents bubbles from entering and blocking the blood vessels.

The specifics of the preparation (solution composition, perfusion pressure, flow rate, temperature, pH) are not subject to investigation in this book and depend on many factors such as the animal species and study objectives. For a review of the Langendorff method see [59].

2.1.2 Optical mapping

Classic electrophysiological techniques for the study of cardiac dynamics employ electrical probes, which need to be in contact with the heart tissue to record an action potential at a specific site. Development of specialized probes with up to several hundred electrodes allowed simultaneous multisite recording and greatly advanced the understanding of the spatio-temporal dynamics in atrial and ventricular arrhythmia. However, next to the usual problems associated with recording of fast-changing signals of small amplitude, namely a trade-off of sampling rate and signal to noise ratio, electrical mapping has a major limitation, which is the inability to accurately record the action potentials during high-voltage shocks. This limits its suitability to study cardiac processes during defibrillation.

In cardiology and related fields, the term *optical mapping* refers to a technique for recording the dynamic propagation of excitation waves in the heart muscle without electrodes. The development of optical probes, i.e. dyes which can change their optical characteristics in dependence on a certain property of their environment, facilitated simple and contactless mapping of said properties on large regions of myocardial tissue at high spatio-temporal resolution – also during defibrillation [32, 67]. A commonly used fluorescent molecule is di-4-ANEPPS [5, 12], which serves as a fast-response probe for millisecond action potential changes of excitable cells, such as cardiomyocytes and neurons. The dye is added into the perfusion solution and thereby distributed throughout the heart tissue. The molecule has a tail which is designed to embed itself into the lipid bilayer of the cell-membrane, in order to sense the trans-membrane voltage. Lipid-bound di-4-ANEPPS has distinct excitation and emission spectrum peaks, centered about 470 nm and 620 nm, respectively [12]. The spectral properties depend strongly on the environment. The potentiometric fluorophore component changes its electronic structure in response to a change of the surrounding electric field, resulting in a change of fluorescence intensity and a shift of the absorption and emission peaks. For optical mapping the excitation and emission light is often limited to the falling edges of these peaks using optical bandpass filters (fig. 2.1b), such that the red-shift results in a depression of fluorescence, which is almost linear to the change of electric field. The base fluorescence intensity of a particular point on the tissue depends on factors such as non-uniform distribution of dye and heterogeneities in local absorption due to vasculature or fatty tissue (fig. 2.1d). Overall, the dye shows only up to 10% fluorescence change per 100 mV (fig. 2.1e), which is why typically sensitive photodiode arrays and cameras with sCMOS (scientific complementary metal-oxide-semiconductor) or EMCCD (electron-multiplying charge-coupled device) image sensors are used for acquisition, that allow digitization at high dynamic resolution.

Another important quantity in cardiac research is intra-cellular calcium concentration, which plays a central role in excitation-contraction coupling [31] (fig. 1.1b). During the course of an action potential, the spark of free calcium initiates the contraction of the muscle cell. The Ca^{2+} concentration can be measured optically, too, e.g. with fluorescent indicator Fluo-4 [26, 38]. Simultaneous measurement of membrane potential and intra-cellular calcium can be achieved by dual-imaging using appropriate dyes, excitation wavelengths, and optical filters [22], as has been shown by Choi and Salama [25] and Omichi et al. [36] with two separate photo diode arrays and CCD cameras, respectively. This approach however requires precise alignment of the imaging sensors in order to get matching field of views. Lee et al. [62] have demonstrated multiparametric imaging using just a single camera with frame-interleaved acquisition by alternating illumination. In this work only single-fluorescence imaging with di-4-ANEPPS is performed, since the focus is on advancing the methodologies towards 3D panoramic imaging of the beating heart. Nonetheless, care was taken that dual imaging could still be easily added at a later stage, using more cameras or frame-interleaved acquisition.

2.1.3 Motion artifacts

Optical mapping of dynamic wave phenomena in cardiac tissue has a major limitation, which is its sensitivity to motion of the sample. Under specific conditions, so-called *motion artifacts* arise in the recorded time series of the fluorescent signals, which can distort the true course of the signals up to complete obfuscation.

One might think that it should be possible to ignore the motion, if it is relatively small compared to the wave length of the excitation waves, especially in case the net movement of the sample over time is zero. However, this is only true if the amplitude of the optical signal is sufficiently large. As mentioned previously, some fluorescent dyes such as di-4-ANEPPS have a small fractional change of only 10 % per 100 mV. This means that the amplitude of the optical signal can be smaller than the typical spatial variation of the base fluorescence, due to inhomogeneous distribution of dye and natural absorption variations in the tissue (figs. 2.1d and 2.1e). When the sample does not move, each pixel of the camera is associated with a specific point of the tissue (pixels P_1, P_2, P_3 and tissue points A, B, C in fig. 2.1b). As the base fluorescence only changes slowly over long periods, the recorded time-series can easily be filtered with detrending algorithms. However, when the sample moves, the pixels

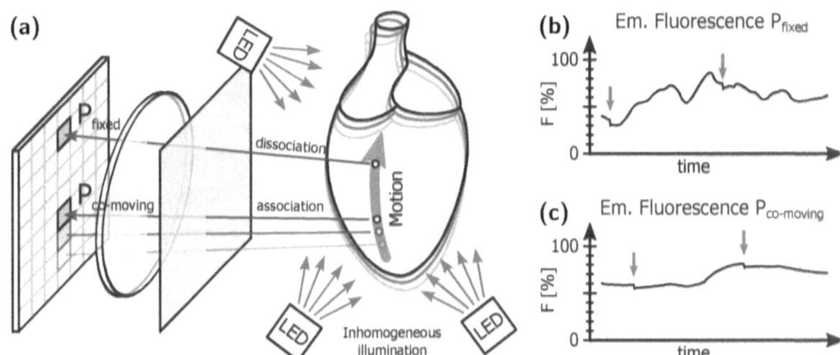

Figure 2.2 – Motion artifacts. (a) Schematic showing two causes for motion artifacts. (b) Emission fluorescence recorded by fixed pixel shows timeseries dominated by spatial variation of base fluorescence, with only minor deflections of action potentials (red arrows). (c) Recorded fluorescence recorded by co-moving pixel shows timeseries dominated by spatial variation of base fluorescence, with only minor deflections of action potentials (red arrows). **Remark:** panels b and c show illustrative artificial data.

and tissue locations become dissociated (P_{fixed} in fig. 2.2a). A single pixel registers light emitted from varying locations on the tissue, which may have a different and unknown base fluorescence. Irregular movements and/or irregular base fluorescence patterns result in irregular motion artifacts, which can hardly be distinguished from action potentials in question. Depending on the fractional change in fluorescence intensity, the signal can be completely obscured by the motion artifacts (fig. 2.2b).

Another contribution to motion artifacts comes from spatially inhomogeneous illumination. It has a similar effect as the spatial tissue variations, since the base amplitude of fluorescence emission linearly depends on the amplitude of excitation.

A simple way to reduce motion artifacts is prevention of the movement in the first place. The chemical compound blebbistatin is a myosin inhibitor which is widely used to suppress the contraction of cardiomyocytes, while the ionic dynamics of action potential formation remain mostly unaffected, compared to other excitation-contraction uncouplers [44]. However, blebbistatin has some negative characteristics, which are photo-instability to blue light, cytotoxicity, low solubility in water, and fluorescence. This makes its experimental application challenging [70]. Furthermore, in a study on rabbit hearts Brack et al. [71] found that "blebbistatin significantly affects cardiac electrophysiology. Its use in optical mapping studies should be treated with caution."

Ratiometric imaging is a different approach that can be used for reduction of motion artifacts, where the tissue movement does not need to be inhibited [11, 27]. Instead, the emission spectrum of the dye is evaluated at two separate wavelength intervals, yielding two signals which have inverse dependence on the dynamic variable but share the same motion artifacts. Thus, in ratio the common artifacts cancel each other, while the dynamic signal remains. Different variants of this approach have been developed, such as excitation ratiometry or emission ratiometry. Additional equipment and methodology are needed, in order to filter and register the images with two cameras. Alternatively the image can be split side-by-side, or frame-by-frame for use with a single camera. Ratiometric imaging does improve the signal quality greatly. It can deal with motion artifacts produced by tissue variations and inhomogeneous illumination. The movement of the tissue itself does not need to be known, and therefore usually also remains unknown. Then, dissociation of tissue location and detector pixel still occurs, which should only be ignored for small movements.

Alternatively, motion tracking techniques offer a twofold benefit. First, the movement of the tissue is recorded and becomes accessible to analysis itself. Second, the association of tissue point and corresponding image location is reestablished, allowing a co-moving analysis of the fluorescence signals ($P_{\text{co-moving}}$ in fig. 2.2a and fig. 2.2c). However, this approach on its own cannot reduce motion artifacts caused by inhomogeneous illumination. Therefore it can be accompanied by ratiometry. The problem of inhomogeneous illumination is discussed in more detail and addressed in chapter 7, where a new purely image-based 3D motion tracking technique is developed, that uses only a single wavelength but also allows to estimate and compensate the effects of inhomogeneous light field.

2.2   Ex vivo panoramic optical mapping setup

Figure 2.3 shows a photograph of the large ex vivo panoramic optical mapping setup of Research Group Biomedical Physics (BMP) at Max Planck Institute for Dynamics and Self-Organization (MPIDS), depicted in the state at the beginning of this book work.[1] The perfusion system can be used for hearts of medium to large size, e.g. rabbit and pig hearts. It consists of a heated reservoir tank for the perfusion solution, pump, additional heat exchanger, and a commercial Langendorff system[2] (bubble trap, pressure regulator with manometer and overflow, aortic block), whose parts have been separated and attached to a custom made aluminum frame on top of an optical table. The aortic block is the part where the heart is connected to, and which allows to inject drugs and monitor flow rate.

While it is possible to have the heart surrounded by air, in our case it is immersed in a perfusion bath, for two reasons. First, disturbing reflections on the wet heart surface and droplets are avoided, which is beneficial for optical mapping. Second, the electrolytic fluid around the heart makes it possible, that electrodes for acquisition of electrocardiogram (ECG) signals or delivering of defibrillation shocks do not need to be in direct contact with the heart tissue and can be placed at some distance, therefore granting unobstructed view onto the epicardium. Usage of a bath is generally beneficial for temperature control, prevention of cell loss on the surface due to dry up, and maintaining of a physiological osmotic pressure. All in

---

[1] There exist also similar smaller setup for panoramic optical mapping of mouse hearts, which is not used in this work.

[2] Isolated Heart Size 9 Type 842, Hugo Sachs Electronic, Harvard Apparatus, Germany

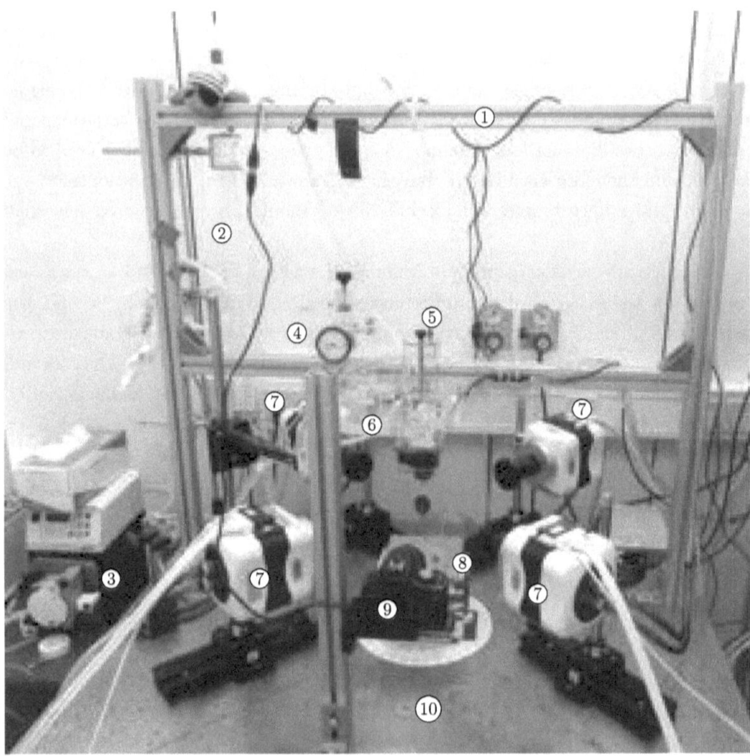

Figure 2.3 – Experimental setup for Langendorff perfusion and optical mapping of rabbit and pig hearts. Photograph taken shortly after beginning of book work. Parts: 1 static aluminum frame, 2 heat exchanger, 3 pump, 4 manometer, 5 bubble trap, 6 pressure regulator and aortic block, 7 four fluorescence cameras on rails, 8 lab jack, 9 geometry camera, 10 optical table. Reservoir, bath and excitation LEDs, and rack with computer and electronic equipment not shown. Motorized rotation stage already installed under aortic block.

all, the tissue remains stable for a longer time in which the heart can be used for experimentation.

A DSLR (digital single-lens reflex) camera[3] with 18–55 mm zoom lens is used for taking high-resolution color images of the heart for subsequent 3D shape reconstruction. This camera will be later on referred to as the *geometry camera*. Panoramic 360° optical mapping can be done with four liquid-cooled EMCCD cameras[4], which have a 128 × 128 px back-illuminated sensor, 16 bit digitization, and up to 530 Hz acquisition rate at full resolution. These cameras will be referred to as *fluorescence cameras* throughout this document. The cameras are arranged on rails around a

---

[3] EOS 500D, Canon Inc., Japan
[4] Evolve 128, Photometrics, Tucson, AZ, USA

central octagonal glass bath. Two sets of prime lenses (fixed focal length photographic lens) are used for rabbit[5] and pig[6] hearts. Excitation of fluorescence in hearts stained with dye di-4-ANEPPS is done with 530(10) nm green LEDs[7], which can be arranged around the bath. They are additionally highpass filtered at a cutoff wavelength of 560 nm [107]. Lenses are equipped with 610 nm red emission filters.

Fluorescence cameras are controlled with software MultiRecorder, which was developed in our group by myself and colleagues for simultaneous acquisition from multiple cameras. Recordings from all cameras are synchronized with common TTL (transistor-transistor logic) trigger signals, which are generated by a function generator or data acquisition device[8]. Delivery of high-energy electrical defibrillation shocks can be done with a bipolar amplifier[9] (max. 100 V, 10 A) or custom-built experimental defibrillator[10] containing a series of capacitors, capable of delivering standard single defibrillation shocks, as well as performing LEAP (low energy anti-fibrillation pacing) defibrillation protocol [54, 63, 90]. The experiment can be centrally operated with software Pulsar, which has been developed in our group for remote control of most devices and software, and logging. A multi-channel amplifier and acquisition system[11] records additional data channels from signal sources such as ECG electrodes, MAP electrode, flow sensor.

2.3 Improvements of experimental setup

The experimental setup has already been used for 3D panoramic optical mapping for some years, as the basic method was established in our group by myself during my Diploma book time. However it has many inadequacies (mostly mechanical stability), which have been addressed as part of this work, to make the setup more stable and usable for frequent experimentation.

For our needs it is vitally important that all mechanical equipment does not move during the course of a perfusion experiment, otherwise carefully obtained camera calibration is lost, because calibration procedure cannot be repeated during the experiment while the heart is attached. This applies to the whole mechanical setup, since the movement of only one component can affect the optical path of one or more cameras. Easy and reliable adjustability of all components greatly affects the time needed for preparation before an experiment. Furthermore, the setup is also used for other experiments where calibration is not needed and flexibility is key.

In the following the improvements to the various components are described. The redesigns comply with these general criteria:

- sturdy construction,

---

[5] Navitar 25mm F0.95, Navitar, Rochester, NY, USA
[6] Navitar TV Lens F0.95/17mm, Navitar, Rochester, NY, USA
[7] LUXEON Rebel green
[8] NI USB-6259, National Instruments Corporation, Austin, TX, USA
[9] BOP 100-10MG, Kepco Inc., USA
[10] Shock Box "Debbie", MPI DS, Göttingen, Germany
[11] BIOPAC MP150 system with STM100C, UIM100C, DA100C & TSD104A, and software Acq-Knowledge, BIOPAC Systems, Inc., Goleta, CA, USA

- easily and finely adjustable without tools,
- adjusted positions must not change when clamped and be resistant to accidental touch,
- repeatability of set positions (if possible).

2.3.1 Bath support stand

Previously, the perfusion system was mounted at a fixed height and the bath was placed on a height-adjustable laboratory support jack (lab jack) below the aortic block. This arrangement poses two problems with respect to the aim of a sturdy and accessible setup. First: In order to access the heart inside the narrow bath, the lab jack has to be lowered. Second: The horizontal position of the bath on the support plate of the lab jack is not defined. Fixing the bath on the plate with adhesive tape or some sort of clamps is complicated and not reliably repeatable. Third: the lab jack is mechanically unstable and the height position not reliably repeatable. To overcome these problems, the lab jack was improved and a new vertical translation frame was constructed.

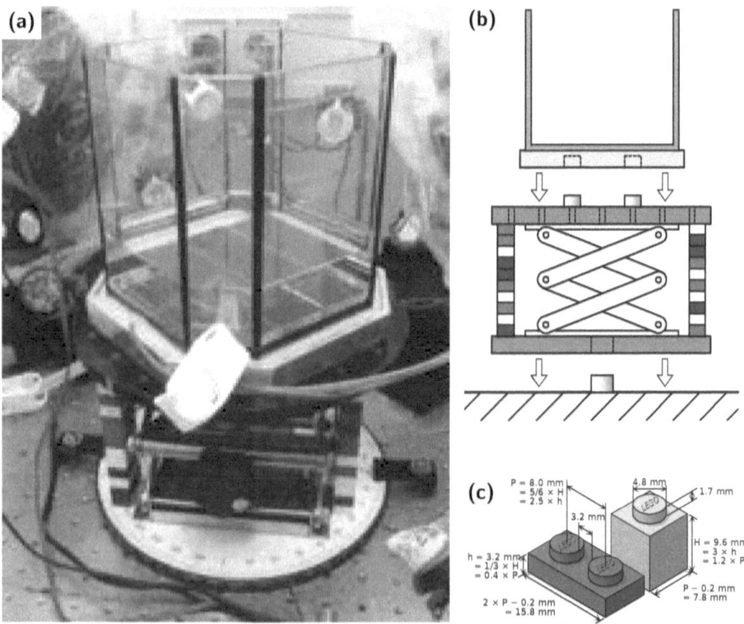

Figure 2.4 – Improved laboratory support jack. (a) Finished device with top and bottom plates and bath mounted on top. (b) Diagram of improved bath and support stand stacked onto optical table. The bath support stand can rotate about a central cap screw head, while the bath is hold in position being stacked on two screw heads. (c) Nominal dimensions of standard Lego® bricks.[12]

Commercially available height-adjustable support frames are mostly either simple lab jacks with the aforementioned disadvantages, or rugged precision stages, typically used for optical applications. The latter are generally well suited for our needs with respect to stability, but they are usually quite expensive and the vertical travel is limited, especially on motorized versions, which would be desirable for precise repeatability. Instead I decided to use the existing lab jack and modify it such that it meets our requirements. Figure 2.4a shows the improved lab jack. The device was extended with two circular plates of black anodized aluminum. The top plate features a regular grid of M4 threaded holes with 25 mm spacing for mounting of the bath and accessories. The bottom plate features a central hole which tightly fits onto the head of a custom-made cylinder head cap screw. The lab jack can be easily placed onto the screw head, which locks its lateral position on the optical table. Additionally this allows the lab jack and bath to be easily rotated around the vertical axis. Petroleum jelly can be used as grease to allow for easy and fine rotation and prevent water (or worse: perfusion solution) to creep in between the bottom plate and the table top. The lab jack is fixed in position with the help of two clamps on the optical table. The bottom plate also features an angular grid, and a linear indicator can be mounted on the optical table.

The lab jack however still lacks of stability and precise repeatability of vertical position. To address this problems an easy solution proved sufficient. Lego® bricks can be stacked between the top and bottom plate on four sides and the lab jack lowered, tightening the screw until the whole jack becomes rigid. The Lego® bricks have a nominal height of $H = 9.6$ mm (excluding the stud), and the height of plates is a third of that value (see fig. 2.4c). Therefore this method allows to repeatedly fix the lab jacks position in multiples of $h = 3.2$ mm.

2.3.2 Perfusion baths

Existing glass baths were equipped with a base plate, which has holes on the underside in positions compatible with a 25 mm raster. The holes tightly fit onto the heads of M4 cap screws, which have been reduced to a diameter of 10 mm. Hereby, the bath can be easily stacked onto the support stand at a defined position without tightening screws or clamps (fig. 2.4b). The bath can also be removed and reliably reattached in the same position.

New baths follow this design and aim to incorporate all components (electrical heating, temperature sensor, outflow) in the base, in order to keep the glass windows unobstructed at all sides (cf. section 8.3.2). For the same reason, level of liquid is not maintained by an overflow pipe inside the bath, but the outflow tube is connected to an external overflow, which additionally can be adjusted in height to set the liquid level.

---

[12]Drawing created by Cmglee, used unchanged under license CC BY-SA 3.0 [115], file source:

2.3.3 Vertical translation frame

In order to have a second option to access the heart in the perfusion bath, besides changing the height of the bath stand, the aluminum frame holding the bubble trap and aortic block was rebuild to be vertically movable (fig. 2.5). Aluminum profiles[13] of size 40 mm × 40 mm and corner brackets were used for the basic strong and rigid construction. Diagonal struts were added to prevent back and forth movement. Linear guide rails[14] and pillow blocks with manual clamps[15] allow for vertical travel of an inner frame of about 450 mm. Pulleys and counterweights balance the load for easy one-hand operation. Rulers on both sides allow precise adjustment of the vertical position. Additional freely movable manual clamps below the sub-frame allow to mark its current position for temporary lifting and exact repositioning.

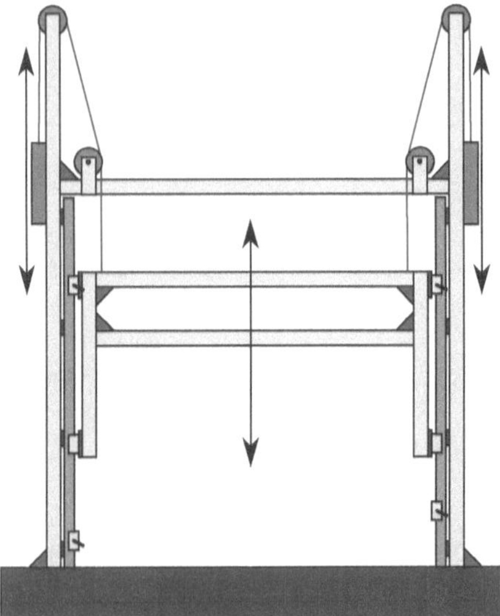

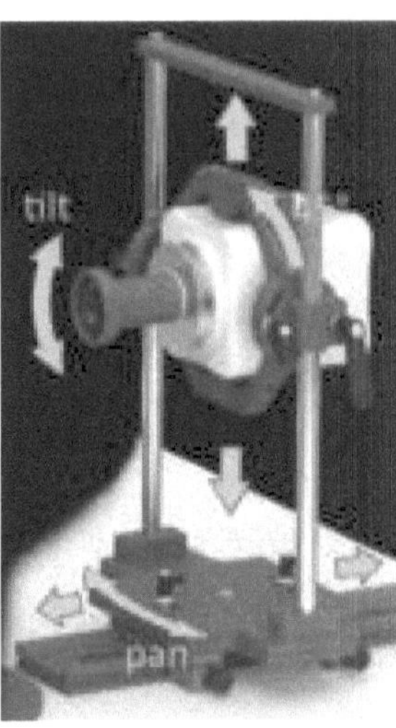

Figure 2.5 – Vertical translation frame. The inner frame can be moved with a lifting system, composed of wires, pulleys, and counterweights (red). Manual clamps (yellow) allow easy fixation at specific heights.

Figure 2.6 – Camera rails with five separate degrees of freedom. All axes (except 45° orientation) can be easily adjusted without tools.

2.3.4 Camera rails with five degrees of freedom

The old camera rails had a mixture of metric and imperial screws of different sizes, which made adjustment tedious and time consuming. More severely, clamps of rotating elements could be overloaded by the load of the quite heavy cameras and

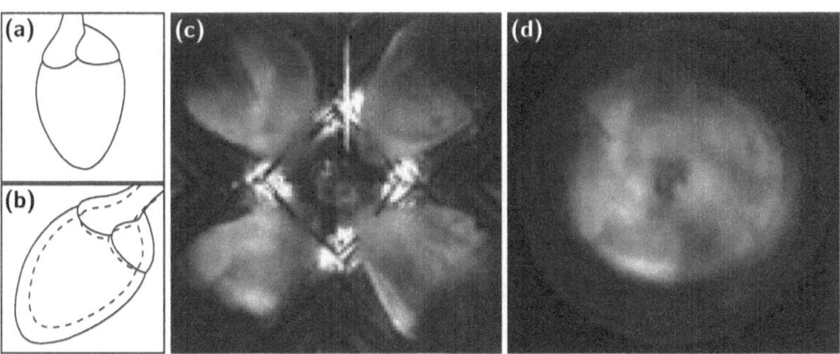

Figure 2.7 – Comparison of camera mounting angles. (a) Best-fit view of heart with camera mounted in usual 0° orientation. (b) Enlarged best-fit view of heart with camera mounted in 45° orientation. (c) Composition of pseudo-continuous polar view of the heart surface from views of four cameras mounted in 45° orientation. (d) Polar texture obtained by projection of camera images onto reconstructed 3D surface mesh.

additional torque of the cables. Furthermore, due to asymmetrical installation of the cameras over the rails, adjusted positions could change during fastening of screws. Therefore the camera rails were completely redesigned with each DOF (degrees of freedom) separately adjustable (fig. 2.6).

Commercial 95 mm wide optical rails and carriers with manual clamps[16] were selected for linear positioning, on top of which custom-build, strong and rigid camera positioners were constructed. Rotation of the whole assembly around the vertical axis (±20°), vertical translation along sliding posts, and rotation around horizontal axis of camera (±45°) can be adjusted separately and fixed with manual clamps (no tools required). Apart from the C-mount for lens/microscope attachment, the Photometrics Evolve 128 cameras have 1/4" mounting holes on four sides, which is good, because the casing is quite 'wobbly' if only one hole is used. Therefore, a camera holder was designed that encompasses the camera and holds it from all four sides. The outer shape of the holder is octagonal, allowing the camera to be rotated around the optical axis and mounted in steps of 45°. In case a camera needs to be retracted to give way for other equipment during the experiment, its position along the rail can be marked and restored with the help of extra carriers.

The possibility to mount the camera rotated about the optical axis allows for better exploitation of the image area, as is illustrated in fig. 2.7. Rotating the camera by 45° and moving it closer to the heart, the image can be zoomed by approximately 20 %, improving light yield (compare figs. 2.7a and 2.7b). Additionally, this camera orientation allows for compact, pseudo-continuous polar view of the heart surface, which is useful for easy and fast preview of the dynamics on the whole heart surface, without having to perform 3D shape reconstruction and image projection to obtain a polar texture map (compare figs. 2.7c and 2.7d).

---

[16]Rail FLS 95, Carrier X 95 - MB, Qioptiq Photonics GmbH & Co. KG, Göttingen, Germany

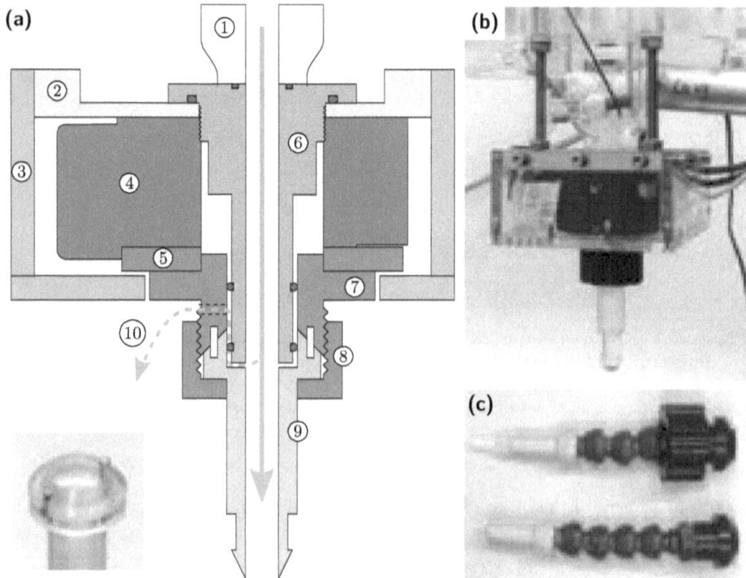

Figure 2.8 – Motorized rotation stage and heart perfusion adapters. (a) Schematic cross section of stepper motor with rotary feed-through: 1 perfusion system's 'aortic block', 2 steel top cover, 3 acrylic glass casing, 4 motor body, 5 rotating stage, 6 feed-through stator, 7 feed-through rotor with conical connector, 8 swivel nut, 9 heart perfusion adapter, 10 safety vent hole pathway. Purple: o-ring seals. Inset: Photo of conic adapter with rotation locking pins. **(b)** Photo of motor assembly mounted under perfusion system. A short perfusion adapter for pig hearts is attached to the motor. **(c)** Heart perfusion adapters for rabbit (top) and pig (bottom) hearts made from articulated tube for better positioning of heart in cameras' fields of view.

2.3.5 Motorized rotation stage

In the old setup the heart could be rotated around the vertical axis for acquisition of multiple photos for 3D reconstruction. However, this had to be done manually, by removing a pin, turning, and inserting the pin to lock the heart in a specific angle. To make the process faster and more accurate, a motorized rotation stage[17] was introduced in the system for computer-controlled rotation of the heart.[18] The perfusion solution is passed through the central opening of the rotation stage (fig. 2.8a). A custom-made rotary feed-through guards the internals of the motor against the liquid. It consists of two parts, i.e. stator and rotor, which are sealed with rubber o-rings. The rotor is attached to the rotating stage and features a conical connector for attachment of different heart adapters, calibration targets, or other equipment. In case the primary o-ring should fail during experiment, a vent hole and secondary o-ring step into the breach and protect the motor from the pressurized perfusion

---

[17]T-RSW60A-KT03, Zaber Technologies Inc., Vancouver, British Columbia,

[18]Canada Work on this project already began before start of this book.

solution. A housing made from acrylic glass guards against solution dripping from above. Perfusion adapters of different lengths and sizes were manufactured for connection of rabbit and pig hearts. The conical shape of the connector and adapter ensures self-centering attachment, while small embedded pins prevent turned attachment with undefined angle. This guarantees, that in position 0° of the motor the local x-axis of an attached calibration target is aligned with the x-axis of the motor, which in turn is aligned with the whole table. In addition to straight perfusion adapters (figs. 2.8a and 2.8b), bendable adapters have been made from articulated tubes, that allow for better positioning of the heart to be in frame of all cameras (fig. 2.8c). These articulated tubes can be adjusted by hand, but are rigid enough to hold the heart in place.

Custom-written software MoteraCtrl (Motor and camera control) was developed for automated rotation and acquisition of images. The application controls the stepper motor and sends trigger signals to the programs controlling the cameras (MultiRecorder and Canon EOS Utility). The number of images, angular increment, exposure time, and pre-exposure delay time can be set. The latter allows for relaxation of swinging motion of the heart after rotation by the motor.

2.3.6 Stabilization frames

Hearts are usually not hanging straight down after connection to the perfusion adapter, because the aorta is a strong artery that exits the left ventricle and makes a hairpin turn at the aortic arch. To keep the heart centered in the field of view of all cameras and prevent unwanted spontaneous drift, stabilization frames were developed, with the aim to push the heart into the desired direction at the apex. The pushing elements cannot be attached to the bottom of the bath, since the heart needs to be free for rotation by the motor for 3D shape reconstruction. Instead, everything has to be attached to the motor itself. Figure 2.9a shows the first revision of such a stabilization frame, which comprises an acrylic glass cross, that slides over the rigid tube of the heart perfusion adapter as can be seen in fig. 2.9b. However, it turned out to be difficult to adjust the holding arms at the bottom during experiment inside the bath. Therefore revision 2 was built, where four long L-shaped arms are directly attached to a more rigid stainless steel cross by metal brackets, that can be adjusted individually (fig. 2.9c). The rods are covered by black shrinking tube, in order to avoid reflexes and make image segmentation more easy. Additionally the tube serves as an insulator, preventing the rod to act as an electrical conductor, which may cause undesired side effects in certain situations.

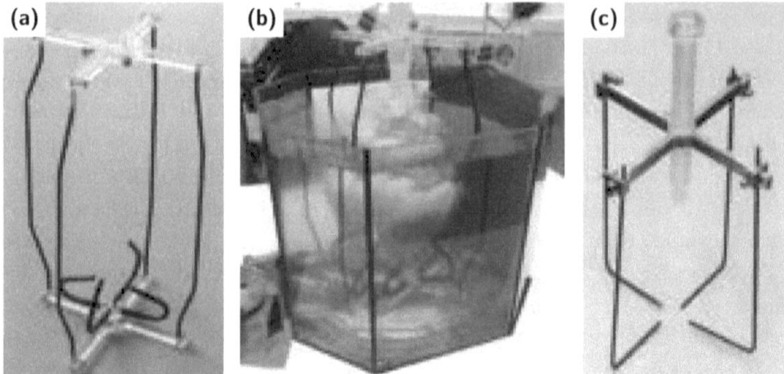

Figure 2.9 – Heart stabilization frames that slide over the rigid tube of the heart perfusion adapter. **(a)** Frame revision 1. **(b)** Stabilized pig heart. **(c)** Frame revision 2.

2.4 Discussion

A photography of the finished improved optical mapping setup is shown in fig. 2.10. The various mechanical improvements make the setup fit for 3D shape reconstruction and 3D panoramic optical mapping. The experiment was moved into another building, which allowed the use of a larger optical table with access from all sides. This greatly benefits the experimental procedures and may even allow the use of more cameras in the future.

Some aspects leave room for future improvement. The new mounts for the fluorescence cameras are very useful, but a little bit bulky due to the large cameras. This leaves very little space when the cameras are placed 45° apart. The current advancements towards smaller, lighter, and cheaper cameras suitable for optical mapping [91] will permit more slender holders in the future and make room for other equipment on the table. The introduction of automated, motorized rotation was necessary for fast and high-quality image acquisition of camera calibration and 3D heart shape reconstruction. This process could be accelerated even more, if the cameras would take images during continuous rotation of heart, which would require proper synchronization of the stepper-motor with the cameras and sufficiently short exposure times. Additionally, in the event of the stepper-motor slipping due to a failing o-ring, currently proper position is lost, because the controller of our stepper-motor model counts the steps from a home position, in order to determine the current angle of the rotation stage. This leads to false estimate of the angular position if the motor is stalled. This problem can be easily solved, because there exist advanced models of the motor with built-in angular encoder and identical dimensions of the rotation stage.

Figure 2.10 – Improved experimental setup for Langendorff perfusion and optical mapping of rabbit and pig hearts.

Chapter 3

Camera calibration

3.1 Introduction

Photogrammetry is the science of making measurements from photographs, which is almost as old as photography itself. Camera calibration is the process of fitting the parameters of a mathematical camera model, such that the mapping between 3D world coordinates and 2D image coordinates of a certain physical camera (lens and sensor) is determined as good as possible. Accurate 3D models of an object can then be reconstructed, e.g. by triangulation of points on the surface of the object with a calibrated stereo-pair of cameras. Full 360° reconstructions can easily be obtained by rotating the object on a turntable in front of a single calibrated camera.

The camera model describes the projection of light from real-world objects onto the imaging sensor, taking into account the optical properties of the lens such as focal length and possible distortions (intrinsic model parameters), as well as the camera's position and orientation in the world coordinate frame (extrinsic parameters). Different camera models exist to describe the various types of lenses, e.g. perspective, fish-eye, or tele-centric. Another aspect to consider is the environment in which the camera is used. For under-water applications, the refraction at the water-glass and glass-air interfaces of the camera's casing can be incorporated into the camera model.

Previously, variants of a common perspective pinhole camera model [7] has been used by others and ourselves for 3D reconstruction of Langendorff-perfused hearts suspended in air [24, 35, 42] and also submerged in perfusion solution within a surrounding bath [46, 51, 97]. The model does incorporate typical distortions of usual lens types. However, this model does not incorporate refraction explicitly. Nonetheless, it can be used for camera calibration with a bath, under the condition that the camera is placed such that its optical axis is perpendicular to the glass window. Then the refraction is effectively absorbed in the camera model's focal length and radial distortion parameters.

In this work, I also chose to continue to use the pinhole model, which will be described in more detail in the next section, since we only use standard lenses and the model proved to be sufficiently accurate. This model is also used and implemented by the software library OpenCV (Open Computer Vision), which provides routines for camera calibration, image undistortion, and much more. Following an introduction of the camera model equations and basic single-camera calibration principles, a multi-camera calibration method will be presented, which utilizes a simple flat calibration pattern attached to the rotation stage. The method employs a novel automatic compensation for possible misalignments, e.g. that the axis of rotation defined by the motorized rotary stage does not properly align with the figure axis of the calibration target. This chapter also explains how to use calibration data in OpenGL (Open Graphics Library) rendering applications. Finally, the multi-camera calibration method will be validated and the suitability of the camera model for the distortion effects of the perfusion bath under non-perpendicular camera placement evaluated.

3.2 Materials and methods

3.2.1 Camera model

The pinhole model's central equation projects a point from 3D world coordinates $\mathbf{x}^{\mathrm{wld}} = (x^{\mathrm{wld}}, y^{\mathrm{wld}}, z^{\mathrm{wld}})$ to 2D image coordinates $\mathbf{x}^{\mathrm{img}} = (x^{\mathrm{img}}, y^{\mathrm{img}})$:

$$s \begin{pmatrix} x^{\mathrm{img}} \\ y^{\mathrm{img}} \\ 1 \end{pmatrix} = \begin{pmatrix} x' \\ y' \\ z' \end{pmatrix} = \underbrace{\begin{pmatrix} f_x & 0 & c_x \\ 0 & f_y & c_y \\ 0 & 0 & 1 \end{pmatrix}}_{K} \left[\underbrace{\begin{pmatrix} r_{11} & r_{12} & r_{13} \\ r_{21} & r_{22} & r_{23} \\ r_{31} & r_{32} & r_{33} \end{pmatrix}}_{R} \begin{pmatrix} x^{\mathrm{wld}} \\ y^{\mathrm{wld}} \\ z^{\mathrm{wld}} \end{pmatrix} + \underbrace{\begin{pmatrix} t_1 \\ t_2 \\ t_3 \end{pmatrix}}_{\mathbf{t}} \right] \quad (3.1)$$

Here, rotation matrix R and translation vector $\mathbf{t}$ compose the camera pose relative to the origin of the world coordinate frame, and are called *extrinsic* parameters. x^{img} and y^{img} are horizontal and vertical image coordinates of the projected point in pixel units with the origin in the upper left corner. The camera matrix K contains the lens-specific *intrinsic* parameters. The point (c_x, c_y) is the principal point of the optical axis, which usually is located near the image center. f_x and f_y are the focal lengths expressed in pixel units. Points (x', y', z') are finally projected to 2D image plane by dividing by $s = z'$ (so-called perspective division).

Equation (3.1) describes the projection for an ideal lens (therefore the term *pinhole* model). Real lenses show nonlinear distortions, which are usually greater for cheap (webcam) lenses. The extended set of equations of the camera model used by OpenCV include radial ($k_{1\ldots6}$), tangential ($p_{1\ldots2}$) and prism ($s_{1\ldots4}$) distortions:

$$\mathbf{x}^{\mathrm{cam}} = R \cdot \mathbf{x}^{\mathrm{wld}} + \mathbf{t} \quad (3.2a)$$

$$x' = x^{\mathrm{cam}} / z^{\mathrm{cam}} \quad (3.2b)$$

$$y' = y^{\mathrm{cam}} / z^{\mathrm{cam}} \quad (3.2c)$$

$$x'' = x' \frac{1 + k_1 r^2 + k_2 r^4 + k_3 r^6}{1 + k_4 r^2 + k_5 r^4 + k_6 r^6} + 2p_1 x' y' + p_2(r^2 + 2x'^2) + s_1 r^2 + s_2 r^4 \qquad (3.2\text{d})$$

$$y'' = y' \frac{1 + k_1 r^2 + k_2 r^4 + k_3 r^6}{1 + k_4 r^2 + k_5 r^4 + k_6 r^6} + p_1(r^2 + 2y'^2) + 2p_2 x' y' + s_3 r^2 + s_4 r^4 \qquad (3.2\text{e})$$

$$\text{where} \quad r^2 = x'^2 + y'^2 \qquad (3.2\text{f})$$

$$x^{\text{img}} = f_x x'' + c_x \qquad (3.2\text{g})$$

$$y^{\text{img}} = f_y y'' + c_y \qquad (3.2\text{h})$$

The camera model (3.2) can be further extended, in order to account for the case of a tilted sensor (Scheimpflug condition), which is characterized by two rotation angles τ_x and τ_y:

$$R(\tau_x, \tau_y) = \begin{pmatrix} \cos(\tau_y) & \sin(\tau_y)\sin(\tau_x) & -\sin(\tau_y)\cos(\tau_x) \\ 0 & \cos(\tau_x) & \sin(\tau_x) \\ \sin(\tau_y) & -\cos(\tau_y)\sin(\tau_x) & \cos(\tau_y)\cos(\tau_x) \end{pmatrix} \qquad (3.3\text{a})$$

$$s \begin{pmatrix} x''' \\ y''' \\ 1 \end{pmatrix} = \begin{pmatrix} R_{33}(\tau_x, \tau_y) & 0 & -R_{13}(\tau_x, \tau_y) \\ 0 & R_{33}(\tau_x, \tau_y) & -R_{23}(\tau_x, \tau_y) \\ 0 & 0 & 1 \end{pmatrix} R(\tau_x, \tau_y) \begin{pmatrix} x'' \\ y'' \\ 1 \end{pmatrix} \qquad (3.3\text{b})$$

$$x^{\text{img}} = f_x x''' + c_x \qquad (3.3\text{c})$$

$$y^{\text{img}} = f_y y''' + c_y \qquad (3.3\text{d})$$

Equation sets (3.2) and (3.3) compose the full camera model provided by Open-CV version 4. Distortion coefficients also belong to the set of intrinsic parameters. OpenCV provides a routine for undistortion of camera images, based on the intrinsic parameters. For undistorted images the simplified camera model (3.1) applies. For more details on the model and distortion, see OpenCV documentation on camera calibration [117].

3.2.2 Single-camera calibration

To help understand our multi-camera calibration method, the standard procedure of single-camera calibration is briefly recapitulated. Estimation of the camera model parameters is achieved by feeding a calibration algorithm with multiple 2D coordinates $\mathbf{f}_i \in \mathbb{R}^2$ of *feature points* in the camera images and corresponding 3D coordinates $\mathbf{o}_i \in \mathbb{R}^3$ called *object points*, where $i = 1, ..., N$ in both cases. The algorithm performs a global Levenberg-Marquardt optimization of the model parameters to minimize the *reprojection error*, which is defined as the sum of squared distances between the observed feature points and the projected object points:

$$e = \sum_{i=1}^{N} |\mathbf{f}_i - \mathbf{p}(\mathbf{o}_i)|^2 \qquad (3.4)$$

Here, $\mathbf{p} : \mathbb{R}^3 \rightarrow \mathbb{R}^2$ is the projection function according to the full camera model (3.3). Different kinds of calibration patterns are available, for which algorithms can automatically detect the feature point positions to sub-pixel accuracy, and corresponding object points can be easily computed in local or global coordinates. Usually, simple regular patterns are used, such as chessboards or circle grids, but other patterns are possible, as will be discussed later.

Estimation of the intrinsic and extrinsic model parameters can be done separately. First, for intrinsic calibration it is sufficient to take a series of photographies of the calibration pattern under different orientations. The calibration target can be hand-held, the specific positions and orientations do not need to be known in this step. However, the algorithm expects a flat pattern and object points have to be specified in local coordinates with $o_{i,z} = 0$. For a good general calibration it is important, that enough different orientations are provided and the full field of view of the camera is covered.

After intrinsic calibration, the extrinsic model parameters (camera pose R, $\mathbf{t}$) can be estimated, given the intrinsic parameters, and another set of feature points and corresponding object points. Now, the 3D object points are not restricted to lie in a plane and can be provided in the desired coordinate system. For example, if the position of the calibration pattern in a global world coordinate frame is known and object points are specified in this system, then the camera pose will be calibrated with respect to the world coordinate origin.

3.2.3 Multi-camera calibration with compensation of misalignment

Extrinsic poses of multiple cameras can easily be calibrated in a common coordinate frame defined by a flat calibration target, provided all cameras can see the calibration pattern simultaneously. However, in our case all cameras look to a central point, making it impossible that the cameras see the same side of the calibration target. Cuboid calibration targets with grid patterns on all four sides have been used to circumvent this problem [24, 35, 46, 97]. The center point of the box defines a common coordinate system. However, presuppositions have to be made or manual input is necessary to tell each camera at which box side(s) it is looking, in order to establish correspondences between identified 2D feature points and respective 3D object points. For automatic identification of the sides, special 3D calibration objects with distinct feature points on all sides could be used.

Here, I present an alternative multi-camera calibration procedure, which uses a simple flat calibration target that can be attached to and rotated by the motorized rotation stage. The automatic rotation allows to take multiple images of the same pattern under different angles, ensuring non-coplanar object points for intrinsic calibration. Figure 3.1 shows two calibration targets of different sizes for pig and rabbit hearts, manufactured in our machine shop. The small pins in the conical connector ensure that the plate is oriented parallel to the table's x-axis in position 0° of the motor. Another advantage of a flat target is, that different calibration pattern films can be glued on both sides of the plate, depending on application needs. We

Figure 3.1 – Calibration targets for different heart sizes. (a) Pig heart calibration target. (b) Rabbit heart calibration target, featuring a large screw for fine vertical positioning and screws for pattern attachment without adhesive tape.

use a small pattern on one side for calibration of the fluorescence cameras, and a large pattern on the other side for the geometry camera, which has a larger field of view. A specific calibration pattern was horizontally mirrored, laser printed on the backside of transparency film, and the print coated with white paint[1]. The pattern was glued onto the plate using double-adhesive tape, while alignment marks on plate and transparency film ensured good centering. The end result is a water and scratch proof calibration target.

The target plate can be programmatically rotated by the rotation stage, resulting in multiple views of the calibration pattern for which the rotation angles ϕ are precisely known. 72 images (views) are photographed every 5° while the target is being rotated 360°. 2D feature points are searched in all images, and intrinsic parameters are estimated for each camera from all views where feature points could be found, as described previously. For extrinsic calibration, the positions of the object points can be readily computed with respect to a global world coordinate frame, which we define to originate on the axis of rotation. Then, extrinsic parameters of each camera can be calibrated not only from a single view, but from all all views where the calibration pattern is facing the respective camera and the feature points can be located, which reduces the error.

The position of the i-th corner of a chessboard pattern with $N = m \times n$ internal corners on the front (+) or back (−) side of the target plate depends on the lattice parameter a (tile size), and a pattern-specific vertical shift Δz with respect to the

---

[1]Without the white paint, air bubbles between the adhesive tape and the transparency film would cause total reflection when the calibration target is submerged under water, which has a negative impact on automatic feature point identification.

target center, and is given in local coordinates by:

$$\mathbf{o}_i^{\text{loc}}(a, \Delta z) = \begin{pmatrix} \pm a \left(l - \frac{n-1}{2}\right) \\ 0 \\ a \left(\frac{m-1}{2} - k\right) + \Delta z \end{pmatrix} \qquad \begin{array}{l} i = kn + l \\ k \in [0, m-1] \\ l \in [0, n-1] \end{array} \tag{3.5}$$

A rotation and translation transformation $(R^{\text{pat}}, \mathbf{t}^{\text{pat}})$ describes the pattern pose, i.e. its alignment in the local co-moving coordinate frame of the rotation stage, while a matrix $R_z(\phi)$ applies the rotation about the z-axis with current motor position ϕ, to yield the object points in world coordinates:

$$\mathbf{o}_i^{\text{wld}} = R_z(\phi) \cdot \left(R^{\text{pat}} \cdot \mathbf{o}_i^{\text{loc}}(a, \Delta z) + \mathbf{t}^{\text{pat}}\right) \tag{3.6}$$

Theoretically the pattern pose is defined by an offset about half the thickness d of the calibration target plate:

$$R^{\text{pat}} = \begin{pmatrix} 1 & 0 & 0 \\ 0 & 1 & 0 \\ 0 & 0 & 1 \end{pmatrix} \qquad \mathbf{t}^{\text{pat}} = \begin{pmatrix} 0 \\ \mp d/2 \\ 0 \end{pmatrix} \tag{3.7}$$

However, true object point locations can deviate from this, due to:

- imprecision during manufacturing of the target plates,
- off-center attachment of the calibration pattern on the target,
- axis of rotation not identical with symmetry axis of calibration target, due to canted attachment to the rotation stage (fig. 3.2a),
- unknown thickness of the transparency film and adhesive tape.

All of these factors contribute to a *misalignment* (ma) of the true object points from the theoretical positions. Although the misalignment is unknown a-priori, it can be considered to be rigid and unchanging during the image acquisition with respect to the co-moving coordinate frame of the rotation stage. Therefore the misalignment can be estimated and compensated by variation of the pattern pose $(R^{\text{pat}}, \mathbf{t}^{\text{pat}})$ using a Levenberg-Marquardt optimization process, which repeats extrinsic calibration until the mean reprojection error of all views becomes minimal (fig. 3.2b). The cost function first computes the object points for all views in world coordinates according to eq. (3.6) with the current set of alignment parameters. Then the camera pose $(R, \mathbf{t})$ is estimated for all views and reprojection errors computed according to eq. (3.4). Finally the mean reprojection error is returned.

Remarks:

During the optimization, the rotation part of the pattern pose is not represented as a 3×3 matrix, because the 9 elements are not independent as a rotation only has 3 DOF. Instead, the rotation is represented as Tait-Bryan[2] angles α, β, and γ, which describe a sequence of three rotations. I chose the convention to perform the

---

[2]For Euler angles there are 12 possible sequences of rotation axes, which are often divided into

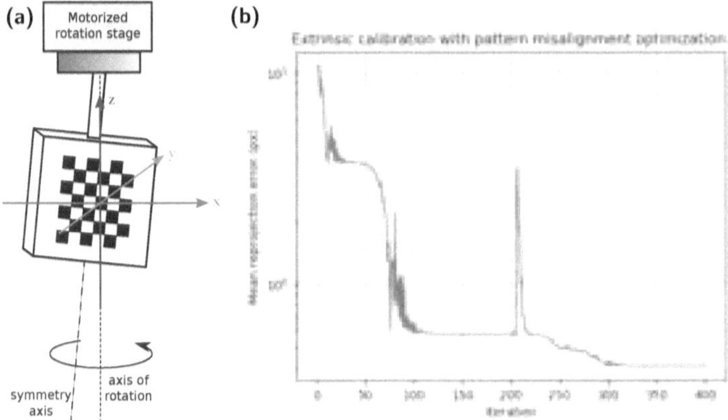

Figure 3.2 – Misalignment of calibration pattern. (a) Schematic of misalignment caused by canted attachment of calibration target to rotation stage (exaggerated). The origin of the world coordinate frame is defined to lie on the axis of rotation. (b) Example of minimization of reprojection error during estimation of calibration pattern misalignment. A Levenberg-Marquardt optimization algorithm is applied in two steps for reasons of robustness (see text), second step starting after iteration 200.

rotations about extrinsic axes in the order x-y-z. This choice allows us to exclude γ from optimization, since the z-rotation does not affect the optimization result. To avoid an ill-defined value, we force the z-rotation to be zero ($\gamma \overset{!}{=} 0$), which ensures that the x-axis of the world coordinate frame coincides with the x-axis of the calibration target in position $\phi = 0°$, as viewed from above.

Similarly, the z-component of the translation is held fixed to be zero ($t_z^{\mathrm{pat}} \overset{!}{=} 0$) during misalignment estimation, because we define the origin of the world coordinate frame to be located at the vertical center of the calibration target on the axis of rotation (compare fig. 3.2a).

Misalignment estimation is performed in two consecutive optimization steps with slightly different cost functions (compare fig. 3.2b). First the camera pose is estimated for each view individually and the resulting rotation and translation vectors averaged before computation of the reprojection error, as convergence with this cost function was more robust. The first optimization result is refined using a second cost function, that directly computes the camera pose using feature and object points from all views, thus avoiding averaging.

Before image acquisition, the mounting of each camera was carefully adjusted, such that the calibration pattern remained fully visible inside the field of view for all rotational positions. The focus of the lenses was adjusted to ensure sharp imaging of the calibration pattern and the heart surface. From here on, all parameters which affected the imaging had to be held fixed, i.e. the position and rotation of the

---

two groups of 6, called Proper Euler angles (first and third rotation about same axis) and Tait-Bryan angles (rotations about three distinct axes).

cameras as well as the focus of the lenses were not changed during the subsequent experiment.

As the geometry camera is set up to have a larger field of view, we use a larger calibration pattern on the back side of the target for intrinsic calibration. Extrinsic calibration is performed for all cameras using the small pattern on the front.

The misalignment is estimated during extrinsic calibration of the high-resolution geometry camera and reused for extrinsic calibration of the low-resolution fluorescence cameras.

3.2.4 Using calibration data in OpenGL

In this work the computer graphics library OpenGL is used not only for visualization (sections 4.2.8 and 5.3) but also for computation purposes using the power of modern GPUs (sections 4.2.7 and 5.2.2). We therefore want to set up a virtual camera in OpenGL to reproduce the projection of a real calibrated camera.

In the following, the OpenGL 3D transformation pipeline is briefly described. It serves the same purpose as the camera model of OpenCV, but while eq. (3.1) just describes the mathematical formula for projection from world to image space, the OpenGL pipeline is broken down into more parts which optimizes the calculations for speed in the context of computer graphics:

$$\boxed{\text{Object space}} \xrightarrow{\text{model transf.}} \boxed{\text{World space}} \xrightarrow{\text{view transf.}} \boxed{\text{Eye space}} \xrightarrow{\text{proj. transf.}} \boxed{\text{Homogen. clip space}} \xrightarrow{\text{homog. division}} \boxed{\text{NDC space}} \xrightarrow{\text{viewport transf.}} \boxed{\text{Window space}}$$

OpenGL uses *homogeneous coordinates* (x, y, z, w) from the field of projective geometry. Using this formalism, model, view, and projection transformations can be expressed as *linear* maps represented by 4×4 matrices. Vertices of 3D objects have to be fed into the pipeline with w set to one: $(x, y, z, 1)$. $w = 0$ represents a point at infinity. Three-dimensional Cartesian coordinates of a point expressed in 4D homogeneous coordinates can be obtained by means of dividing its x, y, and z components by w. Model and view transformations are usually composed of a series of rotations and translations, and are pre-multiplied into a single 4×4 modelview matrix for reasons of speed. The projection transformation to *eye space* (a.k.a. camera space) is usually a perspective or orthogonal transformation. For a perspective camera, a frustum of a pyramid defines the viewing volume. The perspective matrix then projects the vertices from inside the frustum into a rectangular volume. Coordinates are now in *homogeneous clip space*, where vertices with $|x| > |w|$, $|y| > |w|$, or $|z| > |w|$ are laying outside of the visible frustum and can therefore be clipped (triangles crossing the border are cut along the border, and the rest discarded). Clipping is done before the computationally expensive homogeneous division. The projection matrices are designed such that the z-value is copied to the w-component, therefore the perspective division is also performed by dividing by w: $(x, y, z, w) \mapsto \left(\frac{x}{w}, \frac{y}{w}, \frac{z}{w}\right)$. The vertices are now in *normalized device coordinates* (NDC) with all three components in the range $[-1, 1]$. Lastly, the viewport transformation maps x and y

components to *window space*. z-values can be written to a depth-buffer and used for sorting during rendering.

OpenGL uses a right-hand coordinate system for object, world, and eye space, which is convenient for computing. Window space uses the usual image coordinates: origin at the top-left corner, x-axis pointing to the right, and y-axis pointing down. The z-axis is pointing out of the screen towards the viewer, therefore composing a left-handed coordinate system. Usually the viewport transformation maps x from $[-1, 1]$ to $[0, w^{\text{wnd}}]$ and y from $[-1, 1]$ to $[0, h^{\text{wnd}}]$, where w^{wnd} and h^{wnd} correspond to the width and height of the window in pixels. The viewport transformation can alternatively be set up such that the rendered scene does not cover the whole window area, but only a rectangular subregion.

We are now interested in using the model parameters of a calibrated camera (without distortion) in OpenGL. Homogeneous 4×4 projection and view matrices have to be constructed. A good tutorial on how to obtain such matrices from a camera model parameters has been published by Kyle Simek [74]. First, perspective and view matrices can be derived from the calibrated intrinsic camera matrix K and extrinsic pose $(R, \mathbf{t})$ with little modifications:

$$M_{\text{persp}} = \begin{pmatrix} f_x & 0 & -(c_x + 0.5) & 0 \\ 0 & -f_y & -(c_y + 0.5) & 0 \\ 0 & 0 & z_{\text{near}} + z_{\text{far}} & z_{\text{near}} z_{\text{far}} \\ 0 & 0 & -1 & 0 \end{pmatrix} \tag{3.8}$$

$$V_{\text{cal.}} = \begin{pmatrix} r_{11} & r_{12} & r_{13} & t_1 \\ -r_{21} & -r_{22} & -r_{23} & -t_2 \\ -r_{31} & -r_{32} & -r_{33} & -t_3 \\ 0 & 0 & 0 & 1 \end{pmatrix} \tag{3.9}$$

For above result, the different conventions of the coordinates systems used by Open-CV and OpenGL had to be considered, i.e. reversed orientations of y and z-axes, which result in minus signs in the 2nd and 3rd columns and rows, respectively. Additionally, I found that both libraries have a slightly different definition of pixel locations, which is corrected by adding 0.5 to the principal point coordinates. Also note the introduction of the third row in the perspective matrix, which takes care of preserving the depth values between the z-clipping planes. The visibility along the optical axis (negative z-direction) is restricted to the range from z_{near} to z_{far}, which are both absolute numbers and have to be chosen reasonably.

Conversion to normalized device coordinates can be accomplished by multiplying the perspective matrix with the result of an OpenGL helper function `glOrtho(left, right, bottom, top, near, far)`, again observing correct axis conventions:

$$M_{\text{NDC}} = \texttt{glOrtho}(0, \ w^{\text{img}}, \ h^{\text{img}}, \ 0, \ z_{\text{near}}, \ z_{\text{far}}) \tag{3.10}$$

Here, w^{img} and h^{img} are the width and height of the camera images on which the model was initially calibrated, not to be confused with the size of the OpenGL window (w^{wnd}, h^{wnd}). With this, the final projection matrix becomes:

$$P_{\mathrm{cal.}} = M_{\mathrm{NDC}} \cdot M_{\mathrm{persp}} = \begin{pmatrix} \dfrac{2f_x}{w^{\mathrm{img}}} & 0 & 1 - \dfrac{2(c_x + 0.5)}{w^{\mathrm{img}}} & 0 \\ 0 & -\dfrac{2f_y}{h^{\mathrm{img}}} & \dfrac{2(c_y + 0.5)}{h^{\mathrm{img}}} - 1 & 0 \\ 0 & 0 & \dfrac{z_{\mathrm{near}} + z_{\mathrm{far}}}{z_{\mathrm{near}} - z_{\mathrm{far}}} & \dfrac{2 z_{\mathrm{near}} z_{\mathrm{far}}}{z_{\mathrm{near}} - z_{\mathrm{far}}} \\ 0 & 0 & -1 & 0 \end{pmatrix} \tag{3.11}$$

3.3 Validation and results

3.3.1 Validation with rendered images

Validation of the calibration method with misalignment compensation was performed using computer-generated images of the rotating calibration target, as this allowed to precisely control each parameter of the camera model and other imaging aspects and comparison of the calibration result with the ground-truth data. Additionally, these tests ensured that all calibration, conversion, and rendering routines had been implemented correctly. Eighteen different tests with variations of the individual control parameters were conducted. A detailed description of the tests can be found in appendix A, where also full results with all the values of the ground-truth matrices as well as the calibrated camera model parameters and reconstructed OpenGL matrices are listed in tables A.1 to A.20.

Here, an overview of the results is listed in table 3.1. Ground-truth rendering parameters are chosen to simulate the imaging properties of the DSLR (digital single-lens reflex) geometry camera in an experiment with a pig heart. Test cases #0 to #15 each vary a specific parameter of the projection matrix, view matrix, misalignment of the calibration pattern, or distortion coefficients, in order to validate that it can be accurately estimated by the calibration routine. For these cases the reprojection errors after extrinsic calibration are very good, with values of $e_{\mathrm{extr.}} \leq 0.049\,\mathrm{px}$. Original ground-truth images and final renderings based on the calibration are virtually indistinguishable. The only exception is #13, which is not a test like the others, but was included to demonstrate the strength of the effectiveness of the misalignment compensation in comparison to test case #12. Without the estimation of the pattern pose the error increases by three orders of magnitude in this particular case. Test case #14 introduces a set of distortion coefficients that was arbitrarily composed to produce a strong warping effect. Conversely, the second distortion coefficient set (test #15) was not made up, but instead adopted from a calibration result[3] of the real camera (see following section 3.3.2).

Tests #16 and #17 combine the two distortion sets with camera translation (#4) and pattern misalignment (#12). In the first case with arbitrary strong distortions,

---

[3]Distortion coefficients from combination #29, camera $\perp$, medium water, see table B.2.

#	Test description	$e_{\mathrm{intr.}}$ [px]	$e_{\mathrm{extr.}}$ [px]
0	Rendering: target centered, no distortion, no misalignment. Calibration: aspect ratio and principal point fixed.	0.043	0.042
1	As #0, free calibration of principal point.	0.043	0.042
2	As #0, free calibration of principal point and aspect ratio.	0.043	0.042
3	Principal point c_x, c_y not in image center.	0.043	0.042
4	Camera translation $\Delta x = 7$, $\Delta y = -4$.	0.044	0.042
5	Camera rotated 5° about x-axis.	0.049	0.049
6	Camera rotated 5° about y-axis.	0.025	0.023
7	Camera rotated 5° about z-axis.	0.043	0.042
8	Camera rotated 5° about skew axis (1, -4, 8).	0.026	0.024
9	Pattern misalignment 1: translation $\Delta x = -3$, $\Delta y = -0.5$.	0.044	0.043
10	Pattern misalignment 2: rotation -5° about x-axis.	0.034	0.027
11	Pattern misalignment 3: rotation 5° about y-axis.	0.033	0.027
12	Pattern misalignment 4: #9, #10 and #11 combined.	0.037	0.026
13	As above, but without estimation of misalignment.	0.037	25.04
14	Distortion coeff. set 1: strong arbitrary distortion.	0.047	0.044
15	Distortion coeff. set 2: moderate realistic distortion.	0.045	0.043
16	Almost everything: combination of tests #4, #12, #14.	3.535	1.937
17	Almost everything: combination of tests #4, #12, #15.	0.038	0.028
18	Everything: combination of tests #3, #4, #8, #12, #14.	3.638	2.095
19	Everything: combination of tests #3, #4, #8, #12, #15.	0.043	0.030

Table 3.1 – Validation of calibration routine using computer-generated images. Overview of calibration results for various test cases which each vary a specific parameter of the projection matrix, view matrix, misalignment of the calibration pattern, or distortion coefficients. Summary of full results listed in tables A.1 to A.20.

the intrinsic calibration fails, subsequently also causing a large extrinsic reprojection error. There is no apparent problem with corner detection. The OpenCV intrinsic calibration routine fails in this special case for unknown reasons. Further investigation would be needed to find the cause. I did not pursue troubleshooting, since the root of problem seems to be outside of my extrinsic calibration routine. Test case #17 with realistic distortion performed flawlessly. Additional incorporation of off-center principal point (#3) and camera rotation (#8) in test cases #18 and #19 showed the same problem with distortion coefficient set 1 and perfect performance with set 2.

We can conclude that the calibration with misalignment compensation is able to estimate all relevant parameters very well.

3.3.2 Analysis of real distortions caused by water diffraction

To assess the effects of distortions caused by water diffraction, calibration experiments were performed comparing three different test cases, regarding the camera's positioning with respect to the perfusion bath and the medium inside:

1. Optical axis perpendicular to glass window. Medium in bath: air.
2. Optical axis perpendicular to glass window. Medium in bath: water.
3. Optical axis inclined to glass window. Medium in bath: water.

The geometry camera was placed in its usual position for imaging pig hearts, 37.4 cm away from the glass window. For the third test case the camera was lowered by

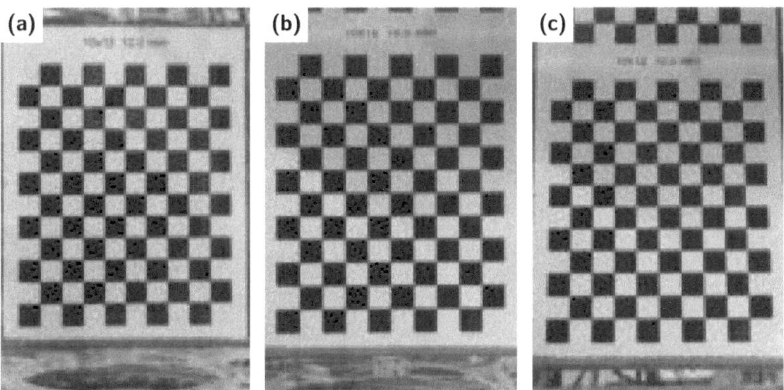

Figure 3.3 – Photographies of the calibration pattern inside the perfusion bath for three test cases. (a) Case 1: Camera positioned with optical axis perpendicular to glass window. Medium air. (b) Case 2: As before, but with water in bath. (c) Case 3: Camera position moved down and pivoted up by 13°.

11 cm and pivoted upwards by approximately 13°. Figure 3.3 shows images of the calibration pattern taken with the geometry camera for these three cases.

The calibration routine of OpenCV is controlled by flags, which allow to fix certain parameters or enable more extended versions of the camera model, eq. (3.3). For the three cases, camera calibration was performed with 64 different flag combinations, in order to see which kind of camera model is best suited and if inclined positioning is feasible. The full results are listed in appendix B, tables B.1 to B.3. An excerpt of the main results is given in table 3.2. For comparison, reprojection errors of the pinhole model without distortions (no dist.) are shown in the first content row. Combination #0 corresponds to a simple camera model, where only tangential and radial distortion coefficients p_1, p_2, k_1, k_2, k_3 are considered. With reprojection errors of 0.427 px and 0.444 px after extrinsic calibration, this simple model achieves quite good results, provided the camera is positioned perpendicular ($\perp$). In the inclined case ($\angle$) the reprojection error of 0.821 px is almost twice as large. However, the error is not exceedingly larger, which is due to the fact that the distortions caused by water and inclined positioning are not very strong, as can be seen in fig. 3.3. When more distortion parameters are considered, the reprojection error becomes generally smaller. Combinations #29 and #30 result in the two lowest reprojection errors for both perpendicular test cases. They are combinations of the rational model with fixed aspect ratio and either the thin prism or tilted sensor (Scheimpflug condition) extensions. In the inclined camera case, #29 also gives a very good result of 0.426 px, but #30 has an surprisingly large error of 3.9 px. Possibly this particular combination of the rational and tilted model distortions is not well suited, or the calibration routine has some instability. On the other hand, combination #6 (tilted model alone) has also quite good error of 0.434 px, similar to that of #29. Other combinations not shown here also yield good calibration results

combination #	fix $k_1 .. k_6 = 0$	fixAspectRatio	fixPrincipalPoint	rationalModel	thinPrismModel	tiltedModel	zeroTangentDist	Extrinsic reprojection error e_{extr} [px]		
								Case 1: $\perp$ air	Case 2: $\perp$ water	Case 3: $\angle$ water
no dist.	✓	✓	✓				✓	1.4592	1.5695	4.1983
0								0.4272	0.4440	0.8209
6						✓		0.3876	0.4185	0.4341
29		✓		✓	✓			0.3746	0.4085	0.4256
30		✓		✓		✓		0.3805	0.4000	3.9381

Table 3.2 – Comparison of calibration performance for three test cases using different combinations of distortion coefficients. Excerpt of full results listed in tables B.1 to B.3. Check marks mean that the respective option of the OpenCV intrinsic camera calibration routine was activated.

for the inclined test case.

We can therefore conclude, that although the camera model does not explicitly model the physical effects of water diffraction, it can be used not only for strictly perpendicular positioning of the camera, but also for inclined viewing. This analysis was done recently and more work needs to be done, to evaluate the angular limits. For all other experiments in this book, the cameras were positioned perpendicular to the bath walls.

3.3.3 Results

Typical calibration results of the geometry and fluorescence cameras are listed in table 3.3 for our use cases of mapping rabbit and pig hearts. The reprojection error values given in pixel units are not directly comparable, due to different imaging conditions (sensor resolution, pixel size, lens type, distance to object). To assess the quality of calibration in more practical terms, the values were also converted to units of millimeters at the respective distance of the camera to the origin of the coordinate system.

	Evolve 128		Canon EOS	
	[px]	[mm]	[px]	[mm]
Rabbit	0.084	0.027	1.4	0.030
Pig	0.046	0.037	0.89	0.033

Table 3.3 – Calibration results. Extrinsic reprojection errors. Values in pixel units are obtained for full frames of $128 \times 128\,\text{px}$ and $3168 \times 4752\,\text{px}$, respectively.

3.4 Discussion

Our approach for multi-camera calibration with a rotating flat 2D target has the advantage that it can be manufactured very easily. The target plate only needs to be flat. The pattern can be printed at high precision with a laser printer. Any imprecision in the manufacturing of the plate, the attachment to the motor, or the alignment of the pattern on the plate is corrected for by the misalignment estimation. Figure 3.2b shows how the estimation of the misalignment reduces the reprojection error by a factor of about 30 for perpendicular test case 2 of the real camera images.

Using this calibration procedure, the camera model can be calibrated to high accuracy of sub-pixel reprojection errors, even though the model equations do not explicitly include water diffraction. A small reprojection error is important, as it directly determines the error in surface reconstruction using Shape from Contour (SfC) method (chapter 4). It becomes even more important for projection of camera images onto the reconstructed mesh, since the depth error δz in direction parallel to the optical axis strongly depends on the viewing angle onto the surface: $\delta z = \delta x \cot \theta$. Here δx is the lateral error and θ is the angle between the surface and the optical axis. While $\cot \theta = 0$ for $\theta = 90$ °, it already reaches a value of 5 for $\theta \approx 11.27°$ and rapidly approaches infinity towards shallower viewing angles at the side of the heart.

For all the experiments in this book, the cameras were positioned such that the optical axis was strictly parallel to the glass normal, to ensure that effects of water diffraction were absorbed into radial distortion parameters of the camera model. Consequently only one camera at a time could be used per window. Geometry and fluorescence cameras had to be exchanged and could not be used simultaneously. The analysis of section 3.3.2 showed that it should be possible to drop this requirement in the future, which will grant more flexibility for positioning of the cameras on the optical table.

The choice of a chessboard pattern for calibration poses another restriction for the experimenter, since the cameras have to be positioned such that the pattern is fully visible. In order to overcome this limitation in the future, other calibration patterns could be used, that allow unequivocal identification of feature points, even if the pattern is partly occluded. One solution to this problem may be cloud-like random calibration patterns introduced by Li et al. [75], which have high and low spatial frequencies that allow calibration from obstructed or detailed views of high and low resolution. Other candidates are calibration patterns composed of multiple unique markers such as (Ch)ArUco boards [79]. However, the low resolution (128 × 128 pixels) of the fluorescence cameras may cause difficulties. Future research has to show if these patterns provide to be useful for our application. Another idea to solve the problem of partially obstructed calibration patterns would be to use a water proof electronic display, maybe an IPx8 certified e-book reader like the PocketBook Ink Pad 3 Pro, in order to automatically adapt the displayed calibration pattern to fit into the field of view of each camera.

A totally different approach for camera calibration is used in a method called Structure from Motion (SfM) [66]. Here, a series of images of an rigid object are photographed, while the camera is moved around the sample. A complex and computationally expensive algorithm then analyses the images and reconstructs the 3D shape of the object, while the camera model and the positions of the views (camera motion) are automatically and simultaneously calibrated. Theoretically, SfM would allow to use any rigid object with strong surface features for camera calibration. The disadvantage of this self-calibration is, that per se it does not give precise absolute units. An intrinsic calibration using some kind of pattern is still advisable. Furthermore, in SfM the coordinate origin is placed randomly, which would have to be corrected for our needs. It seems therefore better to keep our calibration routine, and possibly use the intrinsic and extrinsic calibration data for reconstruction of the heart shape with augmented SfM, as will be discussed in section 4.4.

Chapter 4

Static 3D heart shape reconstruction

4.1   Introduction

A high-quality 3D reconstruction of the epicardial shape is the basis for 3D panoramic optical mapping of static, immobilized hearts (see next chapter 5). The purely image-based surface deformation tracking method presented in chapter 7 also builds upon it. The main parts of a toolchain based on an Shape from Contour (SfC) algorithm were already established in the Research Group Biomedical Physics (BMP) by myself during my diploma book time with help of fellow student and colleague Tariq Baig, following methods used for shape reconstruction of hearts in Langen-dorff preparations published previously [24, 35, 42, 46, 51]. Various modifications and improvements were implemented in our toolchain, but never really published until recently [88, 95]. However, the method sections in those papers had to be quite brief. For this book, the established toolchain was advanced towards automa-tion. Here, more details, reasoning behind, and discussion of the methodological advantages, challenges, and weaknesses of the various steps are presented. An over-all quality validation is performed with a rigid phantom. Alternative image-based reconstruction method Structure from Motion (SfM) will be discussed.

4.2   3D shape reconstruction toolchain

4.2.1   Overview

Figure 4.1 illustrates the step-by-step process of our static heart shape reconstruction method. At first, the series of images of the motionless heart is acquired with the calibrated geometry camera while the heart is being rotated with the motorized rotation stage. The images are pre-processed by a segmentation algorithm to obtain a series of binary masks, which show the hearts contour. Together with the calibrated camera model for each view, a Shape from Contour algorithm computes the 3D shape of the heart. The generated point cloud is further processed with a surface

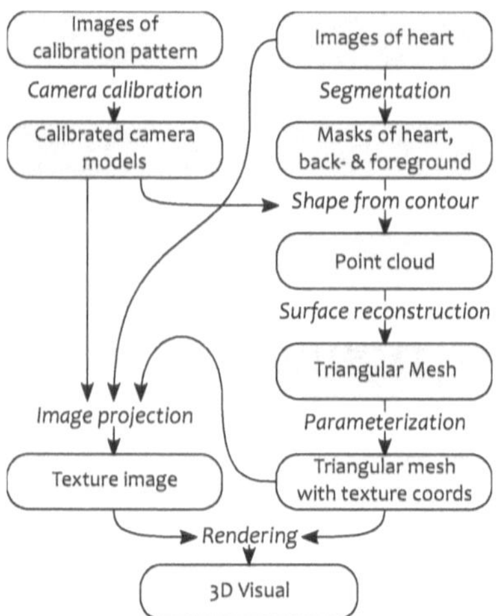

Figure 4.1 – Overview of 3D shape reconstruction toolchain. Starting from multiple images of calibration pattern and the non-moving heart, a series of steps is applied to produce a final pair of triangular mesh and texture image of the whole heart surface.

reconstruction method to generate a triangular mesh of the heart surface. A special spherical parameterization method is used to generate 2D texture coordinates for the mesh. Using the initial images and accompanying calibration data, each image is projected onto the mesh and combined into a 2D texture image of the whole heart surface. The final textured mesh can be visualized using rendering techniques and further used in 3D panoramic optical mapping and motion tracking applications as described in chapters 5 and 7.

4.2.2   Step 1: Image acquisition

Our experiments usually start with calibration of all cameras, after which the actual experiment with the heart is performed. Afterwards the heart is photographed with the geometry camera for subsequent 3D shape reconstruction. For this last step, the heart needs to be immobilized by administration of contraction inhibiting agents such as blebbistatin, if that was not already done for the experiment itself.[1] If the geometry camera was removed after calibration to give way for one of the fluorescence cameras, it has to be carefully repositioned in the original place and assured, that the lens parameters (focus, aperture, zoom) did not change. Also, the camera should be set to fine resolution, low compression, low ISO, and image stabilization turned off. Since we want to use the color information for automatic segmentation, artifacts due to image compression should be avoided. The JPEG algorithm converts RGB

---

[1] If immobilization of the heart only needs to be done at the end for shape reconstruction, then it could be cheaper to use cardioplegia solution instead of blebbistatin. However, the exchange of the whole perfusion solution may move the heart or otherwise affect the imaging.

(red, green, blue) images to YCbCr (luma, blue-yellow chroma, red-green chroma) color model and compresses the color channels more strongly than the luminance. Therefore it is better to set the camera to save images as TIFF or RAW. A non-reflective sheet of uniform color (e.g. green) is put in the bath as background of the heart and white illumination set up. Disturbing objects in the foreground should be removed and reflections avoided. The heart is then rotated with the motor and photographed every 5°. Full rotation and image acquisition takes about 5 min. After each rotation motion, a short delay time has to be awaited for residual swinging of the heart to subside, before the image can be acquired. During this step, it is important to ensure constant perfusion pressure and avoid all other actions that could cause the heart to swell or shrink or otherwise change its shape.

4.2.3 Step 2: Image segmentation

For SfC reconstruction (step 3) the images need to be pre-processed into binary masks of heart and background. For projection of images onto the reconstructed surface (step 6) also image regions showing objects in the foreground should be identified, in order to exclude them from projection. The objective of so-called *image segmentation* is to determine boundaries and divide the images into classes, i.e. different areas which show the heart, the background, and possibly objects in the foreground. It is preferable, that this process runs mostly automatically and unsupervised, since we have a large number of images, typically 72. This is a crucial step in the reconstruction toolchain, since the quality of the final mesh depends on the accuracy of finding the edge of the heart contour. Different approaches have been developed:

Manual segmentation. This is a labor-intensive fall-back solution if every other method fails. Image editing programs typically have specialized tools for this task. For example GIMP (GNU Image Manipulation Program) provides a Scissors Select Tool (select shapes using intelligent edge fitting) and fig. 4.2c shows an example of manual segmentation with this tool. Another GIMP tool is the Foreground Select Tool (select a region containing foreground objects; refine mask by 'drawing' on image, marking the pixels as belonging to foreground or background), which is based on the SIOX (Simple Interactive Object Extraction) method [41, 56]. However, it can only handle two classes (background and foreground) and in the implementation of GIMP the trained classifier cannot be applied to other images automatically.

Thresholding. Segmentation into background and foreground based on image brightness. This is the most simple method and can be easily applied automatically, provided there is good contrast between the heart and the (preferably black) background. It only works if there are no objects other than the heart and background. On the other hand, this method can be even applied to monochrome images of the fluorescence cameras, in cases where the geometry camera cannot be used or the color images are corrupted.

Semi-automatic segmentation. Two methods have been developed for segmentation of color images of the heart in front of a blue or green background.

1. *Filter plugin for GIMP.* This Python script performs segmentation of color images by conversion from RGB to HSL (hue, saturation, lightness) color model. Different thresholds can be set for hue and saturation, in order to distinguish the different components, i.e. heart (hue red, high saturation), background (hue blue/green, high saturation), foreground (low saturation). Gaussian smoothing is applied before final thresholding, in order to remove noise and smooth the contour of the heart. The filter can be interactively applied on individual images, for finding the best set of parameters. Additionally the plugin can called in batch mode, to process all images inside a folder with a fixed parameter set.

2. *Machine learning segmentation.* This method was developed together with my colleagues Alexander Schlemmer, Luca Thiede and Annette Witt. Algorithms from machine learning (ML) library 'scikit-learn' for Python are used for semi-automatic segmentation. First, a small training dataset ($\geq$ 4 images) has to be generated using manual segmentation. Ground-truth classes for background, heart, and foreground were saved as color-coded images (compare fig. 4.2c). Training of the classification algorithm is then performed on a computing cluster. Afterwards it can be applied to all 72 color images.

Results of the different manual and automatic image segmentation methods can be seen in fig. 4.2. Figure 4.2a depicts an original image of the geometry camera showing a pig heart with black stabilization frame in front of blue background. The result of the semi-automatic GIMP segmentation filter plugin can be seen in fig. 4.2b. Thresholds were manually tuned such that contour was best matched at the ventricles. Due to different illumination, the shape of the atria has more noise. Figure 4.2c shows the ground truth classes obtained using manual segmentation with GIMP edge fitting scissors select tool. The result of the ML segmentation method is depicted in fig. 4.2d. It is much better than the GIMP filter and comparable to manual segmentation. The contour of the heart is precise everywhere and the stabilization frame is also resolved very well.

4.2.4 Step 3: Shape from Contour

The three-dimensional shape of the heart is reconstructed from the pairs of 72 silhouette images and accompanying camera calibration data with Shape from Contour method introduced by Niem [15]. The method divides a virtual cuboid volume surrounding the heart into *pillars*, represented by line segments. Initial volume size and spacing of the lines can be adjusted according to the size of the object that should be reconstructed. The segments are then projected to camera image space and truncated where they cross the silhouette contour. This process is iterated for all camera views. Finally, the endpoints of all line segments are returned as a point cloud of the heart surface. Whereas the original method [15] and others only use

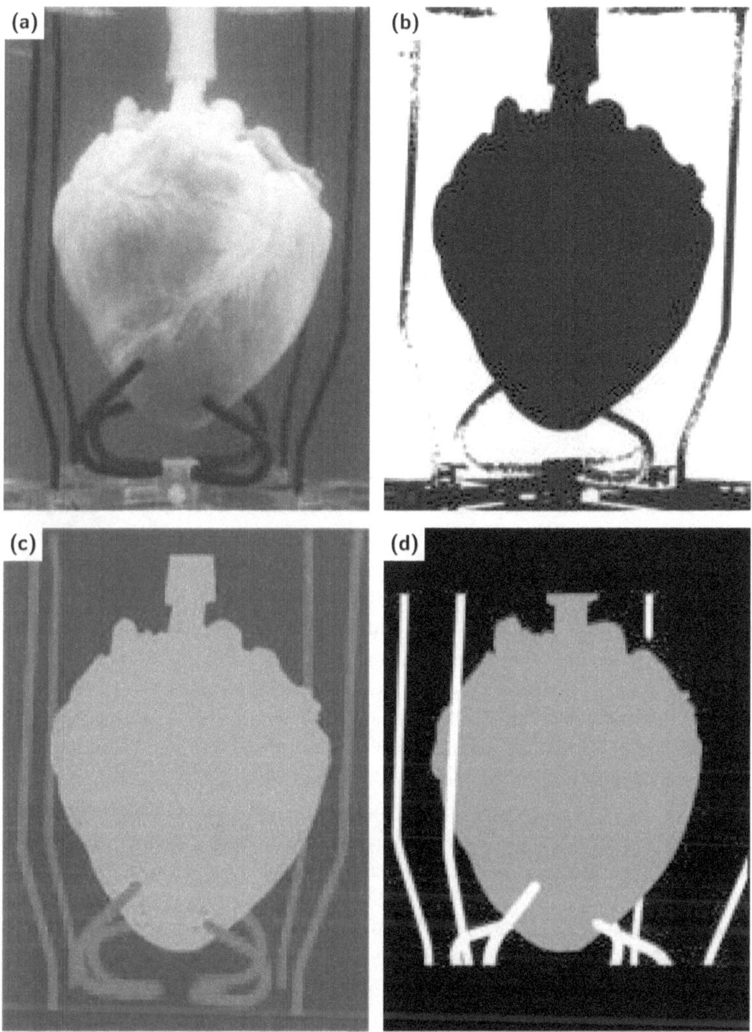

Figure 4.2 – Results of different image segmentation methods. (a) Original image of geometry camera showing pig heart with black stabilization frame in front of blue background. **(b)** Semi-automatic segmentation with own GIMP filter plugin. **(c)** Manual segmentation with GIMP edge fitting scissors select tool. Four out of 72 images are manually processed in this way and provided as ground-truth input for training of the ML classifier. **(d)** Result of automatic ML segmentation applied to image not in training set.

vertical pillars [51], we use lines in x, y, and z-direction, in order to get a more dense point cloud at all sides of the heart (fig. 4.3a).

The lines spacing determines the number of points, whereas the accuracy of the points locations to lie on the actual heart surface is determined by the image resolution and the quality of the segmentation pre-processing. For rabbit hearts a pillar spacing of 0.25 mm is a good choice, whereas for pig hearts a spacing of 0.5 mm to 1.0 mm is sufficient. Because this method bases on a subtractive process, it is important that the silhouette images are free of false holes, which do not correspond to actual holes in the geometry. Therefore segmentation classes of heart and foreground-objects (e.g. stabilization frame) should be combined beforehand.

4.2.5 Step 4: Surface reconstruction

The output of the previous step is a collection of points, which are arranged densely on the heart's surface, but have no actual connectivity relationship. Surface reconstruction is performed using open source software MeshLab [50], which is great for working with such point clouds and triangular meshes, providing tools and routines for manipulation, reconstruction, cleaning, simplification, texturing, and many more. If necessary, the point cloud should be manually cleaned first, e.g. in order to remove points belonging to the stabilization frame. Then a triangulated mesh is reconstructed using Screened Poisson Surface Reconstruction method [73] (fig. 4.3b). To remove small and very thin shaped triangles produced by marching cubes algorithm, the mesh is simplified using Meshlab's Quadric Edge Collapse Decimation filter (fig. 4.3c). This mesh can already be used as the result of step 4.

However, a further *remeshing* with Isoparameterization filter [57] is preferable. This method first builds an abstract domain mesh, which is a coarse almost regular triangular representation of the source mesh. It can then be uniformly resampled at a desired sampling rate, by recursive subdivision of the abstract domain, producing a mesh with very regular triangle sizes and shapes and minimal distortion. As an additional benefit, different sampling rates can produce compatible meshes, where one mesh's vertices are a subset of the other (cf. figs. 4.3d and 4.3e). For example, the sampling rate values $(2, 3, 5, 9, \dots)$ produce a series of compatible meshes. Other combinations such as $(3, 7, 13, \dots)$ are possible. This feature can be useful in case numerical simulations or other calculations should be performed on a simplified version of the reconstructed geometry with one-to-one vertex correspondences, e.g. for non-invasive electrocardiographic imaging (ECGI) mapping (see chapter 9).

4.2.6 Step 5: Surface parameterization

For visualization of the whole epicardium in a single image, a two-dimensional parameterization of the curved heart surface needs to be established, which allows to map each point on the heart to a corresponding location in the image. Cartography provides many different types of map projections of a sphere to a plane, which differ in their strength to preserve or balance local properties such as area size, shape,

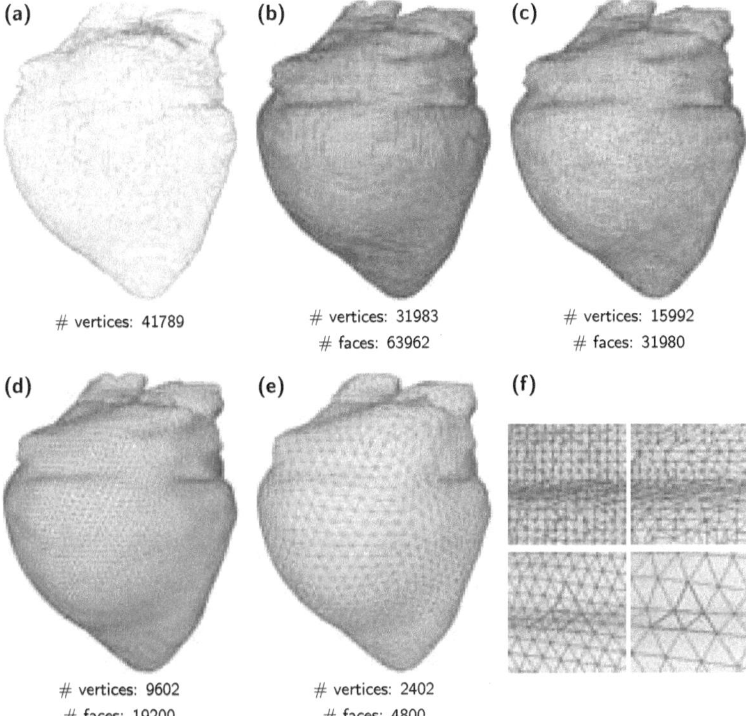

Figure 4.3 – Surface reconstruction and mesh resampling. (a) Point cloud obtained with shape-from-contour method. (b) Triangular mesh generated by Poisson Surface Reconstruction. (c) Mesh simplified by edge collapse to remove small triangles. (d, e) Remeshed by Isoparameterization with sampling rate 9 and 5, respectively. (f) Detail views of panels b to e with corresponding triangles of Isoparameterization highlighted.

or direction.[2] These projections are intended for the case of the almost perfectly spherical earth, where it is sufficient to use latitude and longitude angles in order to uniquely parameterize each point on the surface. However, the heart surface has convoluted parts, most notably at the atria. A simple 2D parameterization which neglects the radial component of 3D spherical coordinates (r, θ, ϕ) can be used for the ventricles, provided the center is chosen appropriately, but does lead to ambiguities at the overhangs of the atria.

Applications in computer graphics (CG) often face the same problem, as 3D models usually have complex shapes, which should be covered with a 2D *texture image*, in order to add visual detail to the often low-resolution triangular meshes. The process of generation of appropriate 2D texture coordinates for the surface of the

---

[2]For a list of map projections see: `https://en.wikipedia.org/wiki/List_of_map_projections`

model is called *UV mapping* in the CG community, as the axes of a texture map are commonly labeled u and v. Different strategies for automatic generation of texture coordinates have been developed. A trivial per-triangle parameterization maps each triangle to a triangular region in texture space, which results in efficient memory consumption, but is impractical for our purposes, as each triangle is disconnected from its neighbors. Atlas based strategies map larger parts of the surface to distinct regions in the texture image, where the dividing seams can be chosen to follow lines separating areas of low curvature in the model. This is good for 3D visualization of complex models, but the texture images itself are difficult to look at and the seams between the regions pose continuity problems for calculations on the images.

Although the heart surface is wrinkled, in our case it is reconstructed as a two-manifold triangular mesh of *genus 0*, i.e. it has no handles and is topologically equivalent to a sphere. This means a spherical parameterization can be established. We use the algorithm of Choi, Lam, and Lui [81], which was developed for the very wrinkled brain surface. The algorithm maps each vertex (x, y, z) of a genus 0 mesh to a unique point on the unit sphere. Then, any traditional map projection of the sphere can be used to generate 2D texture coordinates (u, v).

Previous works on 3D heart reconstruction [51] have used the *equirectangular projection* (fig. 4.4a), which is a convenient choice for visualization, because of its familiar look to popular cylindrical projections used for world maps, like Mercator, Miller, Behrmann, Hobo-Dyer, and Gall-Peters. However, this mapping has two singularities (north and south poles, which are stretched to a line) and one seam (left and right map edges, which belong to the same line on the surface). Computations which operate on the spatial content of the images, such as identification of excitation wave fronts, phase singularities, and conduction velocity, have to consider these special cases imposed by the mapping.

We prefer to use the *azimuthal equidistant projection*, which is a polar representation of the whole spherical surface. It has no seams and only one singularity, which is the outer boundary circle, corresponding to the point on the sphere opposite to the map's central point. The mapping can be oriented such that the central point lies near the apex of the heart and the singularity falls in the electrically inactive region of the aorta. This choice has the advantage, that the whole ventricular (and atrial) surface lies within a connected region free of seams and poles. Abovementioned spatial computations then become very easy to implement.

The 3D triangular mesh is saved as PLY file, which is the standard polygonal file format used by MeshLab and supports texture coordinates as well as user defined properties for vertices and faces in a simple ASCII or binary format.

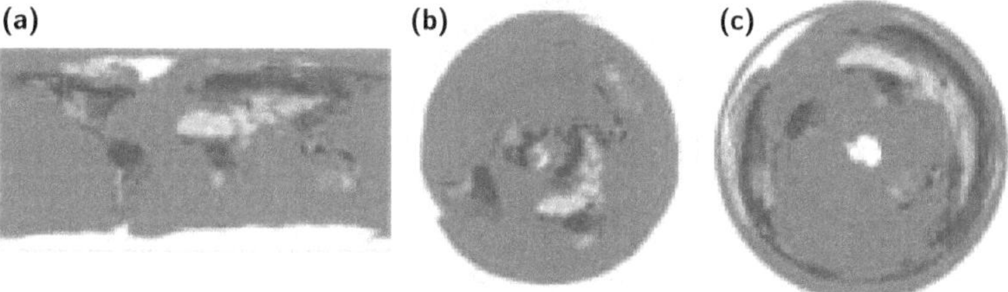

Figure 4.4 – **Different map projections of a sphere.**[3] **(a)** Equirectangular projection (equidistant cylindrical). **(b, c)** Azimuthal equidistant projection, centered at north and south pole, respectively.

4.2.7   Step 6: Texturing – Projection of camera images

Images from calibrated camera views are projected onto the reconstructed 3D mesh and blended into a combined texture image as follows. The implementation uses OpenGL (Open Graphics Library) programmable shaders for fast computation on the GPU (Graphics Processing Unit). First, a texture buffer is set up as rendering target, which has color format RGBA (red, green, blue, alpha) with 32 bit floating point precision for each color component, which is suitable for processing images from the geometry camera as well as fluorescence cameras. For a certain camera view, the triangular mesh is rendered into the texture buffer, taking the texture coordinates of each vertex as its target location. The individual triangles are rendered using the camera image as source texture, where the texture coordinate is computed by projecting the vertices 3D position to image space according to the calibrated camera model as discussed in section 3.2.4. For blending, a weighing factor is computed for each vertex, which is the cosine of the angle between the vertex's normal vector and the optical axis. The weight is set to 0 if the normal is facing away from the camera. The process is repeated for all camera views, and rendered images added up in the buffer. Finally, the texture map is normalized to the original color range, by dividing each pixel's value by the accumulated alpha value. For each camera view, a greyscale mask (generated during image segmentation) can be used to exclude contributions of foreground objects. Note that this method is independent on the choice of method used for generation of texture coordinates. It makes no difference if the triangles are connected or disconnected in texture space. Texture maps can be saved in usual image formats, such as JPEG or PNG, the latter allowing 16bit, transparency, and lossless compression, which removes the negative impact on file size caused by the void areas in the corners of the map.

4.2.8   Step 7: Visualization

The software MeshLab [50] is a good tool for visualization of the static, textured mesh. It can also be used for conversion to other 3D file formats. Many other programs are available for visualization of textured triangular meshes. For dynamic

	True length [mm]		Reconstr. len. [mm]		Absolute & relative error
L_1	81.55(5) 81.35(5) 81.60(5) 81.35(5)	81.46 ± 0.07	81.0745(1) 81.0672(1) 81.0824(1) 81.2303(1)	81.11 ± 0.04	$\Delta_{\mathrm{abs}} = 0.35 \pm 0.08\,\mathrm{mm}$ $\Delta_{\mathrm{rel}} = 0.43 \pm 0.09\,\%$
L_2	81.60(5) 81.45(5) 81.50(5) 81.45(5)	81.500 ± 0.035	81.2393(1) 81.1849(1) 81.2258(1) 81.1852(1)	81.209 ± 0.014	$\Delta_{\mathrm{abs}} = 0.29 \pm 0.04\,\mathrm{mm}$ $\Delta_{\mathrm{rel}} = 0.36 \pm 0.05\,\%$
L_3	91.50(5) 91.45(5) 91.50(5) 91.45(5)	91.475 ± 0.014	91.3577(1) 91.2473(1) 91.2257(1) 91.2292(1)	91.265 ± 0.031	$\Delta_{\mathrm{abs}} = 0.210 \pm 0.034\,\mathrm{mm}$ $\Delta_{\mathrm{rel}} = 0.23 \pm 0.04\,\%$

Table 4.1 – Comparison of dimensions measured on real and reconstructed heart phantom. From four individual measurements the arithmetic mean and standard deviation are computed.

optical mapping data, a 3D visualization tool was written, which will be described in more detail in section 5.3.

4.3 Validation

Due to the many different steps of the 3D shape reconstruction toolchain, which are not all mathematically formulated and may depend on external implementations, a theoretical error analysis of all steps would be very difficult to perform. Instead, a validation of the whole toolchain and evaluation of the reconstruction error was done using a 3D-printed rotation-symmetrical plastic phantom (fig. 4.5a), that has the size of a pig heart and is normally used to align the cameras. The dimensions of the phantom and its reconstruction were measured at three positions (fig. 4.5b):

L_1 Diameter of widest section of the 'atria'.

L_2 Diameter of widest section of the 'ventricles'.

L_3 Distance from a point next to the 'aorta' to the 'apex'.

The real object was measured manually using a sliding caliper, while the reconstructed mesh was first aligned along its symmetry axis and then measured using the measuring tool of MeshLab. For each position, four independent measurements were taken around the phantom with respect to the symmetry axis and subsequently averaged (table 4.1). With absolute errors of 0.35(8) mm or less, the quality of the 3D shape reconstruction toolchain is excellent (for convex shapes).

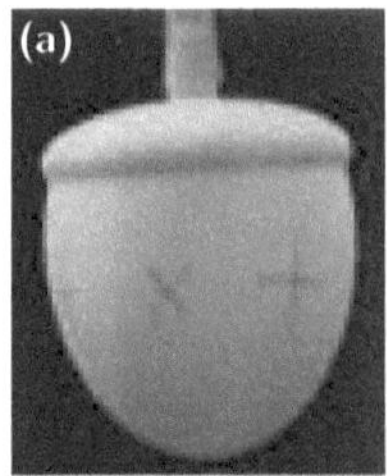

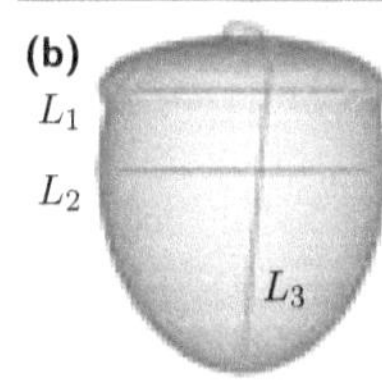

Figure 4.5 – Pig heart phantom. (a) Photograph in front of black background. (b) Reconstructed mesh. Lines indicate measuring positions.

4.4 Alternative approach: Structure from Motion

The 3D shape reconstruction toolchain described in the previous subsections currently requires a lot of manual user control. There exists another image-base reconstruction technique, called Structure from Motion (SfM) [66], which can produce a similar result, but much easier, because it can run automatically and almost unsupervised. SfM is often used for reconstruction of rigid everyday-objects such as buildings, archaeological sites, statues, or devices from a series of images or video frames, where the camera is moved around the sample. The starting point is therefore very similar to our case of a static camera and a rotating heart. Structure from Motion can work on images that have been acquired with such a turntable situation, provided that the background is uniform and does not show features, which would otherwise confuse the algorithm.

Based on theoretical work in the late 1970s [4], and following advances in automatic feature-matching algorithms [6, 8], the Structure from Motion technique was developed by the computer vision community in the 1990s [10, 13, 16]. The SfM algorithm searches the input images for features, performs pairwise matching, identification of outliers, and estimation of object points as well as intrinsic and extrinsic parameters of the camera(s). In an iterative process all of these parameters are mutually refined. If cameras have been calibrated beforehand, intrinsic parameters can be supplied, or alternatively be initialized from Exif (Exchangeable image file format) meta data embedded in images of digital cameras, resulting in a more or less true model size. However, depending on the software used, it may not be possible to also specify extrinsic camera pose parameters, resulting in a random position of the coordinate system origin. After all parameters have been estimated and a coarse model has been reconstructed, a dense reconstruction can be performed.

While Structure from Motion has been on my wish list for a long time, it had to remain there, because we had a working reconstruction process and integration of SfM into our workflow was beyond the scope of this book. It has the potential to drastically simplify our reconstruction toolchain, as it can replace steps 2 to 4, i.e. image segmentation, shape from contour, and surface reconstruction.

4.5 Results

Exemplary results of static heart shape reconstruction of rabbit and pig hearts using our Shape from Contour toolchain are shown in figs. 4.6b and 4.7b. The convex shape of the ventricles is very well resolved, as well as the fine details of the texture map in this region. On the other hand, some concave parts at the atria are 'filled up' and the texture is blurry, indicated by arrows.

A first promising test of the Structure from Motion method has been performed, using the free and open source software Meshroom[4], that was recently released. It has a nice graph based user interface and makes use of a GPU with CUDA support.

---

[4]

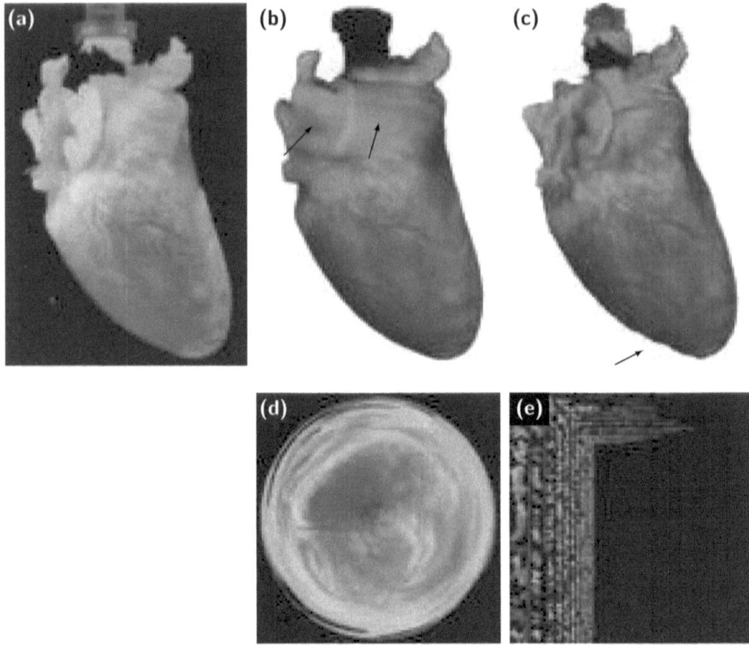

Figure 4.6 – Rabbit heart shape reconstruction results SfC versus SfM.
(a) Original photo of rabbit heart (total 72). (b) 3D mesh obtained with our 3D
shape reconstruction toolchain. (c) 3D mesh reconstructed using software Meshroom.
(d, e) Texture images obtained with our toolchain (spherical parameterization and azimuthal projection) and Meshroom (atlas), respectively.

The reconstruction of the rabbit heart shown in fig. 4.6c was obtained out of the
box using Meshroom's default settings. Comparison with the reconstruction using
the our SfC toolchainshows that the concave regions at the atria are much better
resolved by the SfM method. While the shape of the ventricles is generally well
resolved, some small problems near apex are visible, where the heart tissue curves
away from the camera and does not show strong features. Similar problems arise in
the reconstruction of the pig heart (fig. 4.7c), where the errors at the apex are even
stronger, possibly partly caused by the stabilization frame.

Meshroom does not provide a spherical parameterization for *unwrapping* of the
surface. Only atlas-based methods are available. The texture image produced is
therefore composed of disconnected patches of the original images (fig. 4.6e). To
be suitable for our needs, the mesh would have to be postprocessed by remeshing,
parameterization, and texturing (steps 5 and 6).

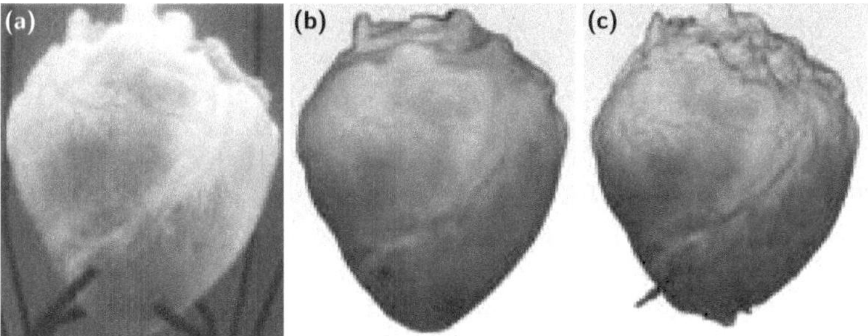

Figure 4.7 – Pig heart shape reconstruction results SfC versus SfM.
(a) Original photo (1 of 72). (b) Reconstruction with our Shape from Contour toolchain.
(c) Reconstruction with Structure from Motion method using Meshroom.

4.6 Discussion

The methods of the static 3D shape reconstruction toolchain produce high-resolution textured mesh of the epicardial heart surface. The relative reconstruction error of 0.43 % represents a lower bound that can be achieved with this methodology under ideal circumstances, as real hearts are not solid and sometimes swell or shrink during the experiment. For comparison, Gloschat et al. [97] performed a similar validation of their shape reconstruction method: "To validate the geometric accuracy, an object of known volume and geometry (i.e. the calibration cuboid) was reconstructed. The reconstructed volume was within 3% of the measured volume of the cuboid."[5]

Many previous works did not use texture images, but instead stored color information on a per-vertex or per-triangle basis [24, 35, 42, 65, 46]. This straight-forward approach however has the drawback, that the resolutions of the mesh and the surface features are directly coupled. In our case, the usage of texture images allows a separation of geometric and color information. This has the benefit that the mesh can be sampled at a resolution which is appropriate to resolve its curved features, while the texture size can be chosen independently according to the resolution of the cameras. The color texture image shown in fig. 4.6d has a size of $1024\,\text{px} \times 1024\,\text{px}$, which results in very fine details. For projection of $128\,\text{px} \times 128\,\text{px}$ images from the four fluorescence cameras, we typically use a texture size of $512\,\text{px} \times 512\,\text{px}$ (cf. section 5.2.2). Textures images can be easily viewed and further processed with image manipulation tools, which are agnostic of the underlying geometry.

The accuracy of the Shape from Contour method depends on the quality of previous image segmentation step. If image is too noisy, the contour masks will not align properly with the real heart surface, and consequently the reconstructed point cloud will be inaccurate.

Many things can go wrong at various points in the acquisition and reconstruction process. Issues we faced:

---

[5]Quote from Gloschat et al. [97, Supplementary Figure 1].

- Heart changing its size due to swelling caused by osmosis, variation in perfusion pressure, or effects of blebbistatin which increase the flow resistance of the vasculature.
- Heart changing its position due to pushing from external stimulation electrodes.
- Blood leaking from heart, caused by clogging during the excision. The blood makes the perfusion solution cloudy, which affects the quality of the image segmentation step.

Such problems can be difficult to compensate for or even unreasonable to continue with the reconstruction procedure, since the resulting errors are too big. All of those problems should not occur in an ideal experiment. But in the real world errors happen, and it is therefore important to further develop methods which are tolerant to device failures and human errors.

In some cases the step-by-step 3D reconstruction toolchain can be of advantage over other integrated methods, since it allows for finer control and more handles for error compensation. For example, if the automatic image segmentation methods have difficulties, one still has the option to switch back to manual segmentation, since the eye/brain is very good at this task in noisy conditions, it just takes more time to do it by hand. As each step in the toolchain is independent from the others (meaning: there are no recursive loops), it can even be replaced by a completely new method more appropriate for the specific problem. A out-of-the-box method like Structure from Motion usually does not allow to compensate for errors which the authors did not anticipate.

While Meshroom does support the OpenCV camera model, currently it only supports 5 distortion parameters. An inclined positioning of the cameras with respect to the glass walls of the bath is therefore not possible (c.f. section 3.3.2). This limitation could be circumvented by calibration of the cameras with our method and supplying undistorted images to Meshroom. Next to the graphical user interface, there exists a scriptable AliceVision/Meshroom toolchain that could be used for customized automation in our workflow. The use of two geometry cameras above each other could probably diminish the shape deviations at the apex. It remains to be seen, if the SfM approach proves itself to be robust enough to replace the SfC toolchain.

The main difficulty of our toolchain is that it currently requires usage of many different own and third-party programs for the various steps. More work needs to be done to give a more streamlined user experience. Meshlab supports scripting for automation, while providing a fallback solution with manual interactive reconstruction using Meshlab's graphical user interface.

Chapter 5

3D panoramic optical mapping

5.1 Introduction

In section 2.2 an experimental Langendorff-setup with multiple cameras arranged for overlapping imaging of the whole heart surface was as presented. On its own, such a system can be readily used for what I call *2D panoramic optical mapping*. Here, the term *2D panoramic* signifies, that although the whole epicardium is observed with multiple views, it can only be analyzed on the basis of the individual 2D projections of the 3D curved heart surface. Correspondences between pixels of overlapping 2D views and with 3D surface points can only be established by eye, and analysis of excitation dynamics is complicated when waves move across image borders. For example, when calculating phase singularities, border effects can falsely increase their actual number, especially at edges of the heart visible in the pictures. Furthermore, identical phase singularities might be counted doubly in the overlapping regions. The calculation of conduction velocity (CV) falsely yields lower values where the surfaces curves away from the camera.

With the addition of camera calibration and 3D surface reconstruction as presented in chapters 3 and 4, *3D panoramic optical mapping* is facilitated, which offers qualitative benefits over 2D panoramic mapping that go beyond the visual appeal of resulting texture maps and 3D visualizations. As the camera images are projected onto the heart mesh and integrated into one texture, the signal-to-noise ratio (SNR) is improved in the overlapping regions, and border effects are eliminated. Furthermore, the same image-based analysis algorithms can be used, that are well-established for 2D (panoramic) optical mapping, i.e. they do not need to be aware of the underlying 3D curved geometry of the heart surface. The choice of single continuous polar texture map of the whole epicardium greatly simplifies the analysis of excitation wave dynamics thereon, since no coordinate discontinuities like seams or poles have to be considered. Tracking and counting of phase singularities is not hindered by borders. Knowledge of the 3D surface geometry allows to correctly compute surface area as well as CV of waves traveling along the curved surface.

5.2 Processing of fluorescence videos

5.2.1 Per-camera preprocessing

Before projection, the original fluorescence camera videos can be preprocessed, e.g. in order to address static image anomalies, like dead or overexposed pixels. We encountered a specific problem of our EMCCD (electron-multiplying charge-coupled device) cameras, which we did not observe in the predecessor model.[1] For synchronous acquisition, the four cameras are fed with a common external trigger signal. Apparently, in this mode the camera hardware can only initiate a frame transfer at fixed intervals, forcing the effective exposure times to be a multiple of about $t_c = 1.4 \times 10^{-5}$ s. At our typical sampling rate of $f_s = 500$ Hz, interference between the external triggering and internal clocking leads to every 5th or 6th frame being slightly shorter exposed and therefore dimmer than the others. Looking at the timeseries of an individual pixel, this small effect can remain undetected within the photon and read-out noise. But for certain computations over larger areas, such as spatial smoothing or averaging over the whole image (*pseudo ECG* computation), the jitter becomes more prominent and disturbing, as it is of a similar magnitude as the small fluorescence signal of the action potential. Fortunately, the affected darker frames can be easily identified in the pseudo ECG and corrected by multiplication with a factor, that can be determined from the first derivative of the pseudo ECG.

Spatial smoothing operations (e.g. binning, or Gauss kernel filtering) should not be performed in preprocessing of the individual cameras, since otherwise pixel values would be spread over unnecessary large areas of the texture image during projection onto the heart geometry where the surface curves away from the camera. Similarly, other procedures that aim to improve the action potential signals, such as temporal low-pass filtering, de-trending, and normalization operations, might perform badly on a single camera with low SNR and should therefore be applied in postprocessing.

5.2.2 Projection onto geometry

The projection of the (preprocessed) camera videos onto the heart surface and combination into a polar texture video is done frame-by-frame with the OpenGL shader as described previously in section 4.2.7. The size of the texture is chosen large enough to prevent undersampling. We found that 512×512 pixels is sufficient to reliably resolve the original 128×128 pixels camera images. In order to avoid loss of information, 16 bit integer fluorescence data are represented as 32 bit floating point values on the GPU. Foreground objects are excluded in the process by masks. The weighted averaging of values from adjacent cameras increases the signal-to-noise ratio in the overlap regions. Parts of the texture image that have missing data or are outside the circular region are set to NaN (Not a Number).

---

[1]EMCCD camera models Cascade 128+ and Evolve 128, Photometrics, Tucson, Arizona, USA.

5.2.3 Postprocessing

The polar texture can be treated like a video coming from a single large camera and accordingly processed in order to prepare the optical mapping signals for further analysis of the excitation dynamics.

Filtering. Depending on the quality of the optical recording, spatial and temporal filtering should be applied for noise reduction. For spatial filtering, I use a Gaussian smoothing kernel, ignoring contributions of pixels with NaN values (c.f. listing C.1). Temporal filtering is done with a Butterworth infinite impulse response (IIR) low-pass filter, with bidirectional processing for zero phase filtering (no group delay). Laughner et al. [67] published a review of processing and analysis of cardiac optical mapping data obtained with potentiometric dyes, in which they very well discuss the important aspects of IIR and finite impulse response (FIR) filter design, in order to avoid introducing filter artifacts that alter the shape of the action potential. However, they only briefly touch the topic of drift removal, mentioning high-pass filtering and polynomial fitting as two possibilities. Such methods are suitable to estimate slowly changing baseline fluorescence, where the smoothing can be calculated over the period of several action potentials. In the case of transitions between different dynamic states, e.g. a slow regular sinus rhythm followed by a fast tachycardia, the different amount of time spent at resting potential leads to vertical shifts in the estimation.

Here I present a filter combination, that aims at selecting only the diastolic parts of the action potential (AP) for estimation of baseline drift (compare fig. 5.1). First, a maximum filter with sliding window length l_{win} is applied to the time series $F(t)$. In order to bridge the gaps in the course of the resting potential, l_{win} should be set to slightly larger than expected duration of an action potential. The result of the maximum filter yields the upper bound of a *smart range*. The lower bound is obtained by application of a minimum filter with the same window length and shifted down by a constant absolute height h_{abs}, or relative height h_{rel} of the upper bound—whichever is larger. These parameters should be chosen according to the amount of noise. The values of the original time series within this smart range are then smoothed with a Gaussian filter of selectable standard deviation σ_{BG}, in order to obtain the estimate of the background fluorescence $F_{\text{BG}}(t)$. Finally, the drift-free fractional change of fluorescence $F_{\text{FC}}(t) = -F(t)/F_{\text{BG}}(t)$ can be computed. Hereby, different intensities due to inhomogeneous illumination are compensated. Depending on the choice of σ_{BG}, short-term motion artifacts can also be eliminated. A Python implementation of the filter using scipy.ndimage is shown in listing C.2.

Figure 5.1 shows an example of the proposed filter applied to recordings of a pig heart, with an maximally observed fraction fluorescence of $\sim$1.3 %. Panels a and b show optical action potentials at two locations with different SNR during pacing. Panel c shows an episode of ventricular fibrillation transitioning to tachycardia after unsuccessful application of LEAP (low energy anti-fibrillation pacing) defibrillation shocks. Results shown in Figure 5.1 were obtained with filter parameter set 1 listed

Table 5.1 – Filter parameter sets.
f_s: camera sampling frequency. σ_{tex}: standard deviation of Gaussian filter for spatial smoothing of texture image. f_{IIR}, N_{IIR}: cutoff frequency and order of temporal Butterworth IIR low-pass filter. l_w: window length of maximum filter. h_{abs}, h_{rel}: absolute and relative height of smart range. σ_{BG}: standard deviation of temporal Gaussian filter for background drift estimation.

Parameter	Set 1 Pig	Set 2 Mouse	Set 3 Mouse
f_s	500 Hz	2000 Hz	2000 Hz
σ_{tex}	4.0 px	1.0 px	1.0 px
f_{IIR}	25 Hz	150 Hz	150 Hz
N_{IIR}	5	5	5
l_{win}	320 ms	40 ms	22.5 ms
h_{abs}	25.0	1.25	1.25
h_{rel}	0.001	0.001	0.001
σ_{BG}	160 ms	20 ms	5 ms

in table 5.1.

Special caution is required when choosing the filter parameters, as otherwise new artifacts can be introduced. The parameters can also be tuned to remove strong motion artifacts. The results shown in fig. 5.2 demonstrate the effect of the filter applied to data[2] of a mouse heart with strong motion artifacts, using two different parameter sets 2 and 3 (see table 5.1). Which one of the two filter parameter sets reproduces action potential shapes that are closer to the true course of the trans-membrane voltage V_m is not easy to decide without an electrical reference measurement. Just judging from the time series shown in fig. 5.2a, one could think there are very little motion artifacts, because the action potential shape looks reasonable for a mouse heart. The difficulty comes due to the fact, that motion artifacts are typically synchronized with the action potential, as they are caused by (residual) contraction. Here, set 3 is assumed to give results closer to the truth. The judgement which part of the recorded timeseries comes from the motion artifact is highly subjective, though. Furthermore, a too small σ_{BG} poses the threat of ironing out important characteristics of the action potential, and introducing distortions in the bridged gaps. This example illustrates the importance of avoiding motion artifacts in the first place, or using techniques such as ratiometry and motion tracking. Due to the strong filtering, the results of figs. 5.2c and 5.2d must be taken with a grain of salt when analyzing the shape of the action potentials, while they are very well suited to study the general course of fibrillation.

In the case of a sufficiently periodic signal, noise can be further reduced by ensemble averaging action potentials over multiple periods. Care must be taken not to average out phenomena with longer periods, such as alternans.

Normalization. In the previous step, the fractional fluorescence was recovered. The signals are therefore freed of the effects of nonuniform dye distribution and illumination. Remaining differences in the amplitude of the fractional fluorescence at different sites can be attributed to inhomogeneities in the tissue, e.g. due to vary-

---

[2]This experimental data was obtained on a similar panoramic optical mapping setup for mouse hearts using cameras MiCAM ULTIMA-L, BrainVision Inc., Japan. The higher SNR comes from the large pixel area and high full well depth of the cameras' image sensor, compared to the Evolve 128 cameras.

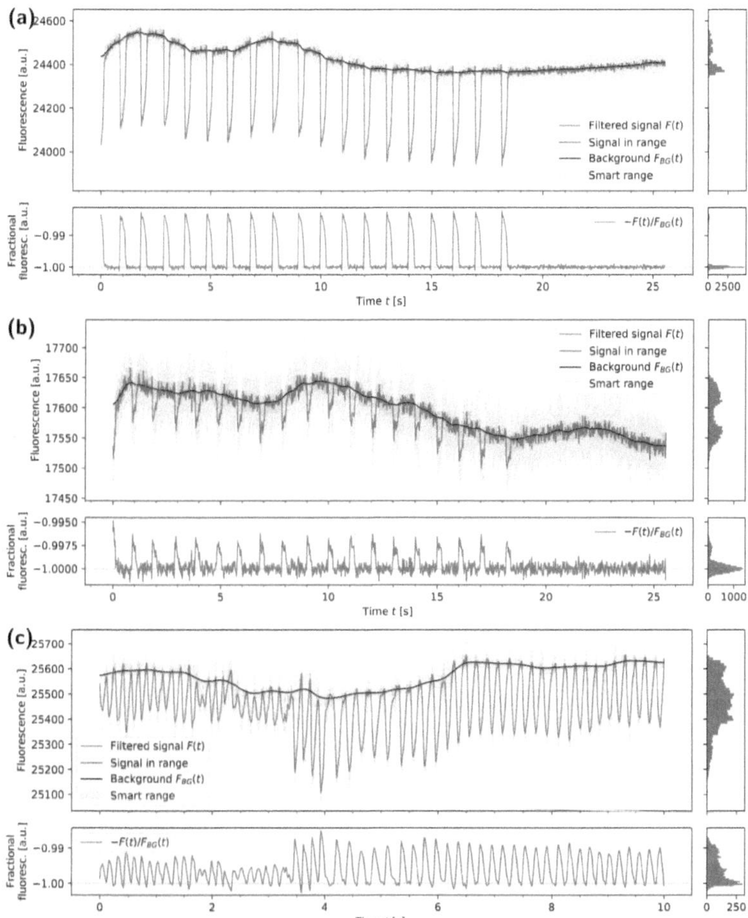

Figure 5.1 – Filtering example pig heart. Original fluorescence signals before (light blue) and after (dark blue) temporal filtering are shown. Parameter set 1 was used for estimation of background fluorescence in smart range (red). On the right, histograms show the distribution of values over the whole time series. **(a, b)** Local pacing. **(c)** Transition from fibrillation to tachycardia after 5 low-energy defibrillation shocks (2.8 s to 4.0 s).

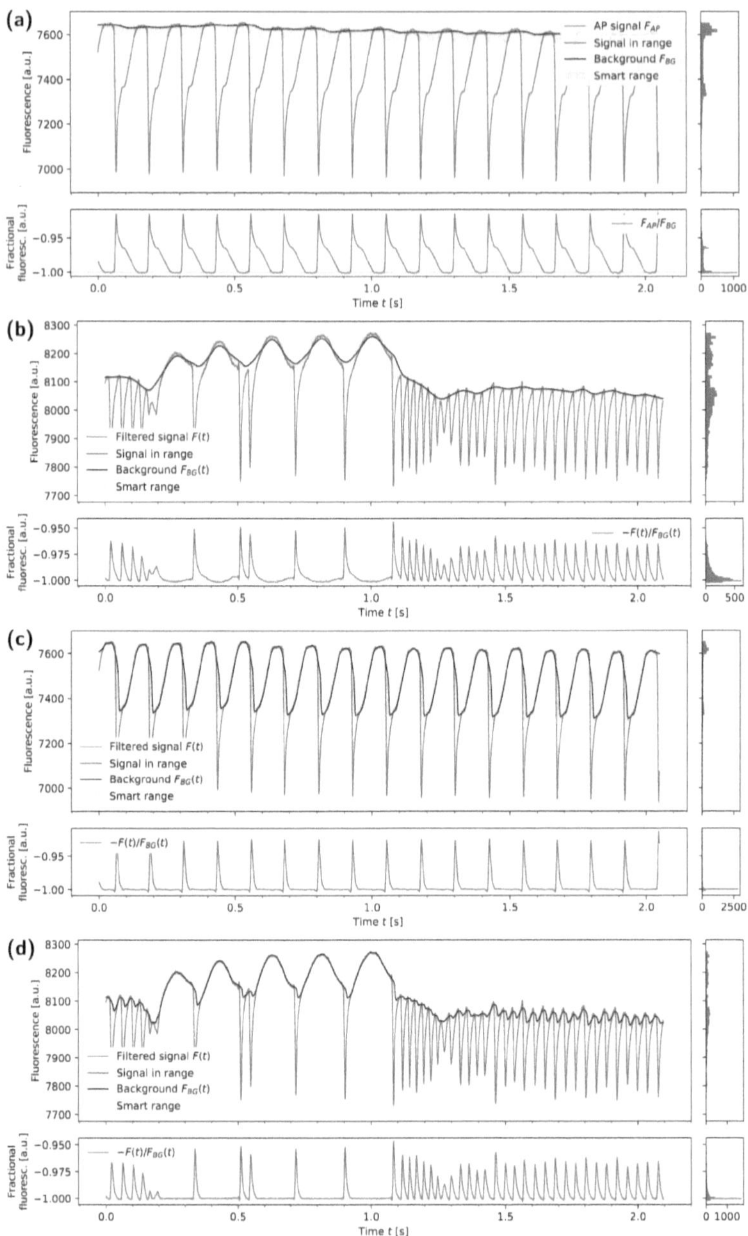

Figure 5.2 Filtering example mouse heart. Comparison of parameter set 2 (panel a and b) and set 3 (panel c and d) with different values of the sliding window length l_{win} and Gaussian smoothing σ_{BG} (see table 5.1). **(a, c)** Sinus rhythm. **(b, d)** Arrythmia.

ing local ratio of electrically active to inactive cells. The standard approach is to normalize the AP to the range 0 to 1, corresponding to the resting and peak potential. However, the identification of minima and maxima is susceptible to residual noise and outliers. Not in all cases ensemble averaging can be used to enhance the SNR. Additionally, we do not want to over-amplify time series of regions that show a quiescent flat line void of action potentials. Time series with a too low signal to noise ratio should be detected and excluded, too. So far, I have not found a good established method addressing this problem.

Here, I present a normalization technique, that builds upon an approach which I already used in my diploma book [69]. The resting potential can be easily identified as the location of the main peak in the histogram of the time series $F_{\mathrm{fr}}(t)$ (see right column in fig. 5.1). For regular activity, a second smaller peak can be seen at the location of the AP plateau. While there is no reason that the AP values strictly follow a specific distribution, I noticed that the sum of two opposing, appropriately shifted and scaled log-normal distributions can be used to fit the course of the histograms quite well for different levels of noise:

$$y(x) = c_1 \cdot f_{\mathcal{LN}}\big(x - s_1;\ \mu_1,\ \sigma_1\big) + c_2 \cdot f_{\mathcal{LN}}\big(s_2 - x;\ \mu_2,\ \sigma_2\big) \tag{5.1}$$

Here, $f_{\mathcal{LN}}(x;\ \mu,\ \sigma)$ is the probability density function with of a log-normal distribution with parameters $\mu \in \mathbb{R}$ and $\sigma \in \mathbb{R}$, $\sigma > 0$, with its support extended to $x \leq 0$:

$$f_{\mathcal{LN}}(x;\ \mu,\ \sigma) = \begin{cases} \dfrac{1}{\sqrt{2\pi}\sigma x} \exp\left(-\dfrac{(\ln x - \mu)^2}{2\sigma^2}\right) & x > 0 \\ 0 & x \leq 0 \end{cases} \tag{5.2}$$

Instead of using a histogram, the time series distribution for each pixel is first approximated smoothly with Gaussian kernel density estimation (KDE). From my experience, choosing the bandwidth with Scott's rule and a factor of 0.5 to 1.0, yields good results. Then, initial values for parameters c_1, s_1, μ_1, σ_1, c_2, s_2, μ_2 and σ_2 are estimated and eq. (5.1) fitted to the data with a non-linear least squares curve fitting routine. The modes of the individual log-normal functions are then computed and stored as base and peak values for later normalization (fig. 5.3a). In order to prevent over-amplification of quiescent and low-SNR regions, a time series is rejected if its standard deviation is below a set value, and base and peak values for this pixel are set to NaN. Pixels can also be rejected, in case the bilognormal distribution does not fit well enough. After all pixels have been analyzed, the Gaussian filter ignoring NaNs is applied to the base and peak fields for smoothing with interpolation of rejected pixel regions. Finally normalization to the range 0 to 1 is performed:

$$F_{\mathrm{N}}(t) = \frac{F_{\mathrm{FC}}(t) - F_{\mathrm{base}}}{F_{\mathrm{peak}} - F_{\mathrm{base}}} \tag{5.3}$$

Recordings containing episodes of fibrillation can be normalized by restricting the KDE computation and curve fitting to time spans of more regular activity (fig. 5.3b).

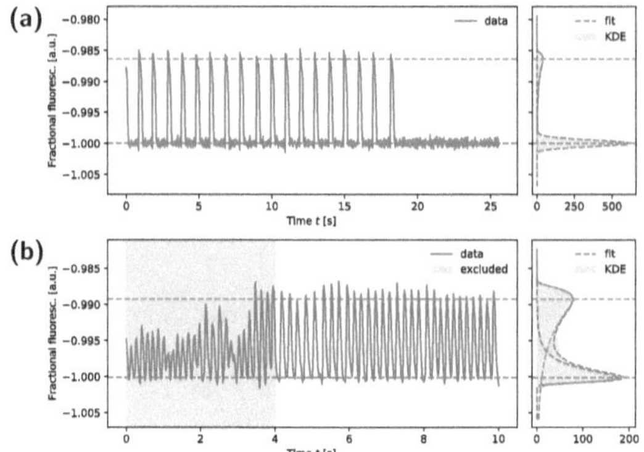

Figure 5.3 – Normalization. Determination of base and peak potentials in noisy optical action potentials by fitting double log-normal functions to distribution of kernel density estimation (KDE). **(a)** Regular action potentials during pacing. **(b)** Exclusion of episode of ventricular fibrillation and restriction of KDE computation to following tachycardia.

Alternatively, smoothed base and peak fields determined for sinus rhythm can be stored in maps and reused for normalization of arrhythmic recordings.

5.3 Visualization

While many programs can be used for visualization of static textured meshes, not all software can deal with transparent or animated texture maps, or other interactive features helpful for our purposes. A plenty of filtering, analysis, and visualization routines for excitable media were already implemented in the software project PythonAnalyser developed in our research group. Due to this circumstance, I decided to port my old C++ code to Python and and write a 3D visualizer plugin for PythonAnalyser. Building upon the library VisPy[3] [76, 94] modern OpenGL (Open Graphics Library) shaders are used for implementation of fast 3D triangular mesh visualization. Multiple data layers can be displayed on the heart geometry at once, e.g. static heart surface color texture, partly-transparent dynamic excitation waves, and annotations. Normalized data can be used as its own alpha map for partly-transparent overlay (fig. 5.4b). Different color maps can be applied to the fluorescence data. Camera model data can be loaded to view the scene from the position of the calibrated cameras (see also section 3.2.4).

---

[3]VisPy, Python library for interactive scientific visualization:

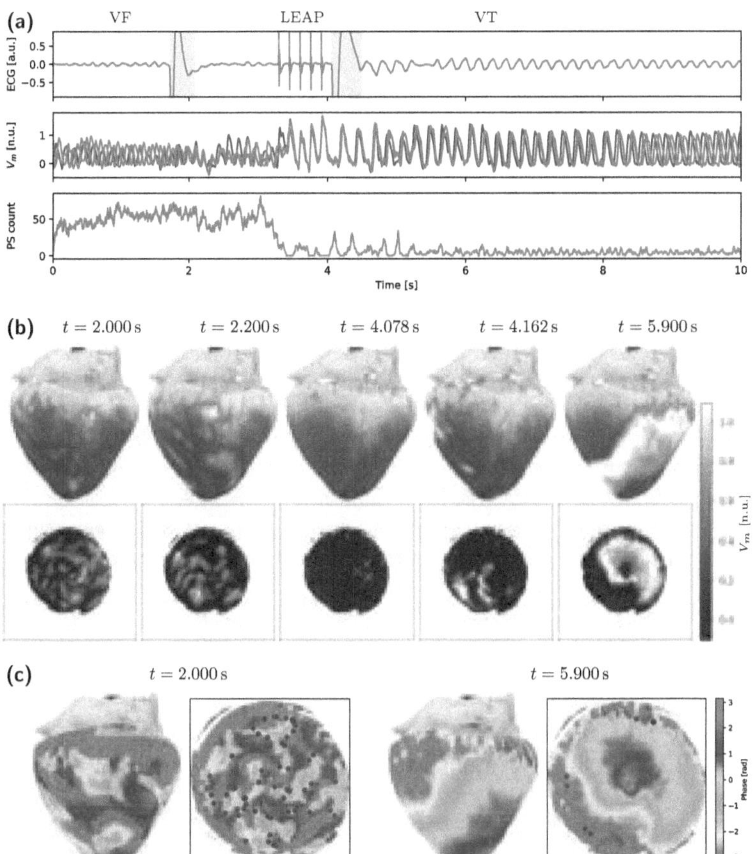

Figure 5.4 – 3D panoramic optical mapping of arrhythmic activity. The pig heart in this example transitions from ventricular fibrillation (VF) to tachycardia (VT) due to an unsuccessful defibrillation attempt with five LEAP (low energy anti-fibrillation pacing) shocks. VT stable for about 40 s, after which it deteriorates into VF again. **(a)** Top: Electrocardiogram (ECG) recording. Artifacts (red time spans) caused by switching of security relays can be ignored. Middle: Normalized optical action potential (OAP) of of three selected neighboring points (compare panel b). Bottom: Number of phase singularities (PS). **(b)** 3D visualization of heart surface mesh with color texture and partial-transparent overlay of normalized optical action potentials (OAPs). **(c)** Phase maps in 3D and 2D (with phase singularities) during ventricular fibrillation and tachycardia, respectively.

5.4 Analysis of excitation wave dynamics

In the following some standard 1D and 2D analysis methods for excitable media in general and cardiac optical mapping signals in particular are briefly introduced and discussed within the context of 3D panoramic optical mapping. Mostly, they can be applied to the video texture with little to no change.

5.4.1 Pre-computation of geometrical properties

The heart surface is reconstructed as an unstructured triangular mesh, where each vertex $\beta = (x, y, z)$ has an associated texture coordinate $\mathbf{a} = (u, v)$. Due to the irregular connectivity, it is useful to pre-compute some geometrical properties over the domain of the texture coordinates and store them in maps of the same size as the fluorescence texture map:

- **Face index map:** The index of the face that contains the central point of the pixel, useful for fast lookup.
- **Position map:** The 3D surface location corresponding to the center of the pixel, useful for simple 3D visualization.
- **Surface area map:** The surface area covered by each pixel. Hereby the surface area of a region of interest in the texture can be easily computed by summation over the area map values of the selected pixels. The map can also be used to normalize histograms that are computed from other maps showing the distribution of certain properties over the heart surface.
- **Map of pushforward tensors:** $T_j^i = \frac{\partial a^i}{\partial \beta^j}$, with $i \in \{1, 2\}$ and $j \in \{1, 2, 3\}$. These tensors are needed for computation of apparent 3D CV on the curved surface (section 5.4.6).

5.4.2 Action potential morphology

Detection of upstrokes in the OAP time series is a central analysis method. As the upstrokes are very fast and pronounced they can be identified quite easily, also in noisy data. Upstrokes or activation times can be defined as the time points with highest slope. Similarly, the slower downstrokes can be identified as the largest local minima of the derivative in between the upstrokes. For calculation of action potential duration (APD) often another definition is used that searches for threshold crossings of the membrane potential at a specific level. For example, APD_{90} is defined as the duration of the action potential at $90\,\%$ repolarization. The duration of the diastolic interval (DI) is defined accordingly. This allows the study of restitution and alternans phenomena appearing in cardiac tissue at short DI [3, 14, 34, 52, 93].

5.4.3 Phase map

The cardiac action potential is a stereotypical excursion of the cell's membrane potential in response to an excitatory stimulus. This can be interpreted as an oscillatory process, to which one would like to assign a phase, not only during regular sinus rhythm, but also during fibrillation. While in simple numerical models of excitable media and cardiomyocytes [9, 18, 48] a phase angle can be directly defined from two state variables (usually the membrane potential and a *refractory* variable serving as the restoring force), the biological action potential is a product of many different ionic currents (c.f. section 1.2.2). Furthermore, in optical mapping often only the membrane potential is measured. In this case, different methods exist to derive an instantaneous phase.

In nonlinear time series analysis [33], a technique called *delay embedding* allows to reconstruct the high-dimensional phase space of a dynamical system using multiple delayed versions $f(t + i\,\tau_d)$, $i \in \{0, 1, \ldots, D_e - 1\}$ of the single original time series, where τ_d is the lag and D_e denotes the number of embedding dimensions. For cardiac signals an embedding into two-dimensional space is sufficient to unfold the general dynamics, provided τ_d is chosen appropriately, according to the shortest expected period of heart rhythm. The signal $f(t)$ should be detrended and oscillating around zero. Then the trajectory $\mathbf{f}(t)$ in this reconstructed phase space more or less rotates around the origin, allowing to compute a phase angle:

$$\phi(t) = \mathrm{atan2}\big(f(t + \tau_d),\ f(t)\big) \tag{5.4}$$

Similarly, following [19] a phase of $f(t)$ can be computed from its complex-valued analytic signal $f_a(t)$, that is defined using the Hilbert transform $\mathcal{H}$, which imparts a phase shift of $-90°$ to every Fourier component of the function $f(t)$:

$$\mathcal{H}f(t) = \frac{1}{\pi} \int_{-\infty}^{\infty} \frac{f(\tau)}{t - \tau}\, d\tau \tag{5.5}$$

$$f_a(t) = f(t) + i\,\mathcal{H}f(t) = \hat{f}_a(t) \cdot e^{i\,\phi(t)} \tag{5.6}$$

$$\phi(t) = \mathrm{atan2}\big(f(t),\ \mathcal{H}f(t)\big) \tag{5.7}$$

In both cases, in order to obtain a signal $f(t)$ that oscillates around zero, often the mean is subtracted for reasons of simplicity. In our case, we have gone through great lengths in order to detrend and normalize the OAP signal. Therefore, instead of using the spatially varying mean, it is more meaningful to subtract a constant value $c \approx 0.25$ that is the same all over the heart: $f(t) = F_N(t) - c$.

In addition to isochronal maps (activation maps), visualization of the instantaneous phase for each point of the heart surface in a phase map is useful for inspection of the heart dynamics.

5.4.4 Phase singularities

In the phase map, certain locations where the phase is ill-defined are of particular interest. Identification of these phase singularities (PSs) yields the pivoting points of spiral waves (fig. 5.4c). A single large spiral wave is thought to be associated with ventricular tachycardia (VT). Wave breakup generates two new counter-rotating spiral wavelets, which can interfere with each other, cause additional breakup and lead to the chaotic activation patterns observed in ventricular fibrillation (VF). Note that phase singularities visible on the heart surface are end points of one-dimensional *phase singularity filaments*, that lie at the core of three-dimensional scroll waves which extend into the thick tissue [49, 82, 95].

Numerical identification of singularities in phase maps can be done by integrating the phase ϕ over a small loop $C(s)$:

$$n = \oint_C \nabla\phi \, \mathrm{d}s \tag{5.8}$$

If the result does not vanish but is close to $\pm 2\pi$, the encircled region contains a phase singularity of positive or negative *charge*, which determines the spiral's direction of rotation.

Another method for identification of phase singularities searches for the points where the lines of wave fronts and wave backs connect [21]. Tracking of the singularities over time can be challenging, due to noise. Berg [107] discusses different methods for calculation of phase, PS identification, and PS tracking.

5.4.5   Activation map

Activation maps show for each location on the heart surface the duration from a specified start time to the next activation (upstroke). Alternatively for live analysis, a running activation map can be computed showing the elapsed time since the last activation until present time. Activation maps are often used to visualize the activation spread through the heart after application of a local stimulus (fig. 5.5) or a global electrical shock. The duration until the full tissue is activated (or e.g. 95 % thereof) is a useful quantity to assess the effectiveness of defibrillation shocks in relation to their pulse energy [63]. The size of activated surface area can be computed by integrating the values of the pre-computed surface area map over the activated pixels in the texture.

5.4.6   Conduction velocity

Conduction velocity is an important measure in cardiac research, as it influences robustness of propagation and is linked to arrhythmia development [108]. Bayly et al. [17] introduced an automated method for estimation of conduction velocity and direction from epicardial mapping data via local fitting of polynomial surfaces to activation maps. This was later extended to estimate 3D CV vector fields within the myocardium using plunge needles [23]. Optical mapping only allows to study the heart surface. Excitation wave patterns visible on the epicardium are only a 2D curved cross section of the 3D dynamics that evolve within the bounds of the muscle tissue volume. Therefore, only *apparent* conduction velocities can be computed from epicardial mapping data. In cases where the direction of wave propagation can be considered to be tangent to the surface (such as quasi-2D dynamics in thin tissue, or wave spread from local epicardial stimulation in thick tissue), the apparent velocity is a realistic estimate of the true CV. When this condition is not fulfilled, i.e. the excitation wave front emerges from within or enters into the tissue, then the velocity on the surface can appear arbitrarily higher than the actual speed of the wave front in 3D. Furthermore, for correct CV computation in optical mapping, the curvature of the surface in 3D space and the projection to 2D image coordinates need to be considered. In 2D optical mapping experiments of whole hearts that do not provide this knowledge, apparent CV is systematically underestimated towards the sides of the heart, where the surface curves away from the camera. Sung, Omens, and McCulloch [29] used a gross 3D model of the heart surface fitted from orthogonal views and a parametric model-based method to estimate conduction velocity from activation times obtained with optical mapping. However, determination of activation times for complex spatio-temporal activity such as fibrillation is inherently ambiguous. Mourad and Nash [45] formulated an explicit expression of the conduction velocity, that uses spatio-temporal gradients of any scalar field for that a particular isovalue defines the electrical excitation wavefronts. This scalar field can be a map of activation times, but also detrended voltage or phase are possible. The latter two are more suitable, as they are also well defined during VF. For estimation of apparent CV, they use a 2D parameterization of the heart surface $\mathbf{S}(u,v) = \left(S_x(u,v),\, S_y(u,v),\, S_z(u,v)\right)$. In our case, we obtain such a parameterization of the triangular surface mesh in the form of texture coordinates for each vertex during 3D surface reconstruction.

Building upon the result of [45], my former colleague Tariq Baig-Meininghaus reformulated the equations into matrix notation and separated time-independent geometric from dynamic parts:[4]

$$\dot{\boldsymbol{\beta}} = -\partial_t f \frac{T^{\mathsf{T}} \overset{\mathbf{a}}{\nabla} f}{\overset{\mathbf{a}}{\nabla} f^{\mathsf{T}} T\, T^{\mathsf{T}} \overset{\mathbf{a}}{\nabla} f} \tag{5.9}$$

Here, $f(\mathbf{a}, t)$ is the scalar field whose isosurface is of interest, $\dot{\boldsymbol{\beta}}$ is the instantaneous (conduction) velocity in 3D coordinates, and $T_j^i = \frac{\partial a^i}{\partial \beta^j}$ the pre-computed pushforward tensor (see section 5.4.1).

If one already has computed an activation map $t_{\mathrm{act}}(u, v)$, then eq. (5.9) becomes:

$$\dot{\boldsymbol{\beta}} = -\frac{T^{\mathsf{T}} g}{g^{\mathsf{T}} (T\, T^{\mathsf{T}})\, g} \qquad \text{with } g = \overset{\mathbf{a}}{\nabla} t_{\mathrm{act}} \tag{5.10}$$

Here the minus sign vanishes, in case the activation time is defined reversely [45].

The pre-computation of the pushforward tensors on the static geometry, allows to compute the 3D conduction velocity from activation maps or other scalar fields directly in the texture domain, essentially making it a 2D analysis method.

---

[4]Personal communication with Tariq Baig-Meininghaus, see appendix D.

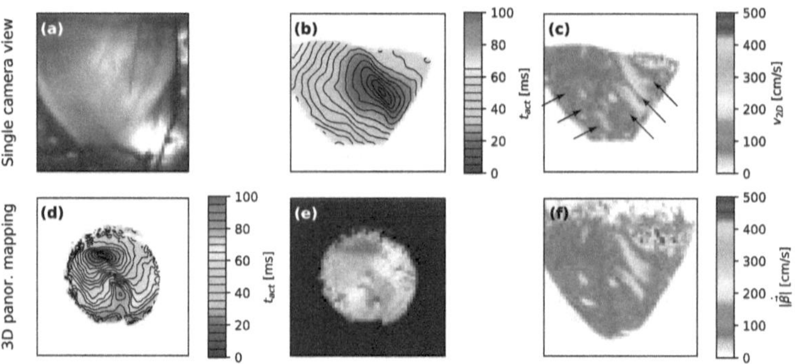

Figure 5.5 – Computation of Conduction Velocity in 2D versus 3D. Example
of excitation spread from local stimulation on pig heart. Action potentials ensemble
averaged over nine periods. (**a**) Camera image showing stimulation electrode. (**b**) 2D
activation map created from single camera. (**c**) In-plane apparent conduction velocity
(CV) computed from (b). CV values are underestimated towards the sides of the heart
(arrows). (**d**) Polar activation map of ventricular surface. (**e**) Polar map of apparent
CV, computed from (d) and pre-computed pushforward tensors with eq. (5.10), showing
color-coded 3D CV vectors. (**f**) Visualization of 3D mesh with absolute value of 3D CV,
shown from the same camera view as top row.

5.5   Example results

The example in fig. 5.4 shows 3D panoramic optical mapping of arrhythmias. Initially, the pig heart was in a state of ventricular fibrillation. At some point, a series of
five low-energy LEAP defibrillation shocks is applied, which successively excites and
synchronizes larger parts of the heart tissue until the whole observable ventricular
surface returns to resting state after activation by the last shock ($t = 4.078\,\mathrm{s}$). Following, excitation waves break through around the apex. The likely cause is residual
ventricular activity, that survived somewhere inside the thick myocardium (e.g. in
the septum). In the following, a sustained tachycardia develops, where excitation
emerges at the apex, spreads outward in a growing circular wave front towards the
atria, disappears at the base, and finally reemerges at the apex, over and over again.
This VT episode lasts for about 40 s, until it deteriorates back into VF. See also
supplementary movie 1 (appendix E).

Figure 5.5 exemplifies the benefit of performing 3D panoramic optical mapping
(bottom row) for the calculation of apparent conduction velocity over 2D single-view
analysis (top row). The excitation waves originate from the tip of a stimulation
electrode and spread outward, as can be seen in the activation maps in panels b
and d. The absolute values of the 2D conduction velocities calculated for the single
camera are systematically smaller at the sides (arrows in panel c), compared to the
absolute values of the 3D CV vectors (panel f) that were calculated with eq. (5.10).

5.6 Discussion

The procedures shown here for a motionless static heart can also be applied to a contracting heart. The only difference is in the projection step, where 3D motion tracking data of the mesh has to be supplied. The image based analysis methods of the texture videos then essentially operate in a co-moving fashion.

The filtering and normalization methods described in section 5.2.3 present solutions for smart detrending, baseline and peak detection which I have not found in the literature so far. However, a more thorough experimental testing of my proposed filters and comparison with other published methods remains to be done. One candidate would be a spatio-temporal Gaussian filter recently published by Pollnow et al. [101], whose parameters were automatically adapted for low SNR signals.

For spatial smoothing operations in postprocessing, a Gaussian filter with a kernel of fixed shape was used uniformly for the whole texture image(s). This leads to *non-uniform* filtering in the surface domain. Instead, it would be better to perform spatial smoothing with an adapted kernel for each pixel of the texture. The principal axes of a two-dimensional Gaussian kernel can be aligned to the local spherical coordinate axes of the texture map and accordingly stretched. In order to speed up the computations, a kernel map can be precomputed, similar to the surface area map. Analogous methods can be applied to the path integration for phase singularity detection.

Part II

Multimodal extensions

Chapter 6

High-resolution CT scanning

6.1 Introduction

With optical shape reconstruction methods, only the geometry of the outer heart surface can be recorded. In situ 4D ultrasound acquisition (chapter 8) can also capture the internal anatomy, but only with limited resolution. It is to our advantage to have in-house access to a CT (computed tomography) scanning machine[1] (fig. 6.1). The device allows fully automated acquisition of samples at resolutions down to the nanometer scale. This enables 2D and 3D metrology applications in industry and science, e.g. in manufacturing to compare of the shape of a produced part with the CAD drawings. For us the device offers the possibility to measure the intramural heart geometry at very high resolution after the experiment.

6.2 Materials and Methods

6.2.1 Computed Tomography

The basic principle of CT is the three-dimensional volumetric reconstruction of a sample from many X-ray projections. In cone beam scanning, X-rays are produced by collimating an electron beam onto a metal target, from which the high-energy photons travel towards a detector screen. In between, a sample is placed on a rotary table, which casts a shadow that reveals its interior structure, since the X-rays are not completely absorbed but only attenuated. The amount of attenuation depends on the density of the sample material and the energy of the photons. As the sample rotates, a large number of 2D images are collected by the detector. The projections are processed to reconstruct a 3D volume image of the internal density distribution of the sample.

The nanotom is a very versatile CT device, as it allows fully automated acquisition and motorized positioning of the sample and detector (fig. 6.1b). By varying of the relative distances between the target, sample, and detector, different magnifi-

---

[1]Phoenix|x-ray nanotom, GE Sensing & Inspection Technologies GmbH, Wunstorf, Germany; usage courtesy of Prof. Dr. Stephan Herminghaus.

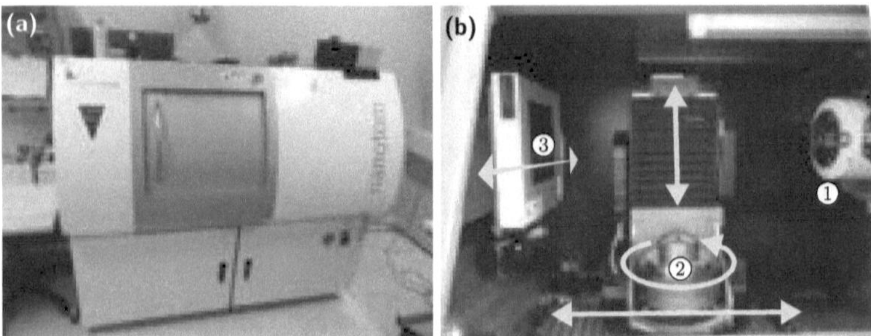

Figure 6.1 – Nanotom CT scanning machine. (a) Exterior view. (b) Interior view. 1: Electron beam target (X-ray source). 2: Sample rotation and translation stage. 3: Detector screen.

cations and resolutions are possible. To extend the field of view, the rotation stage can be moved up & down and the detector screen back & forth.

6.2.2 Preservation of heart tissue

After an ex-vivo Langendorff experiment, the heart was preserved using a histological tissue fixative[2]. The heart was first rinsed, by injecting a portion of the solution retrogradely through the vasculature through the aorta, and then completely immersed in the fixative for at least 24 hours. To prevent deformation by contact with the container walls, the heart was suspended in a floating position with the help of a thread through the lid. The fixation process creates cross-links between the proteins of the cells, making the whole tissue stiffer and preventing its disintegration.

6.2.3 CT scanner sample containers

Sealed cylindrical containers made of acrylic glass were built for the CT scanning of preserved rabbit and pig hearts (fig. 6.2a), which can be mounted to the chuck of the scanner's rotation stage. Since the available space inside the CT scanner is limited, the diameter of the large sample container is not much larger than the width of a pig heart. The heart can be mounted hanging inside the container, similar to the Langendorff experiment. To avoid deformation, the attachment point on the lid of the container can be changed radially, so that the large and asymmetric pig hearts can hang freely without touching the walls of the cylinder.

6.2.4 CT scanning and reconstruction

Before insertion into the container, the remaining fixative solution was shaken out of the heart cavities, which increased the contrast between the tissue and surrounding air in the CT scan. To prevent oscillating movements, soft foamed material was used,

---

[2]Roti-Histofix, acid free (pH 7), phosphate-buffered formaldehyde solution 4%, Carl Roth GmbH, Karlsruhe, Germany.

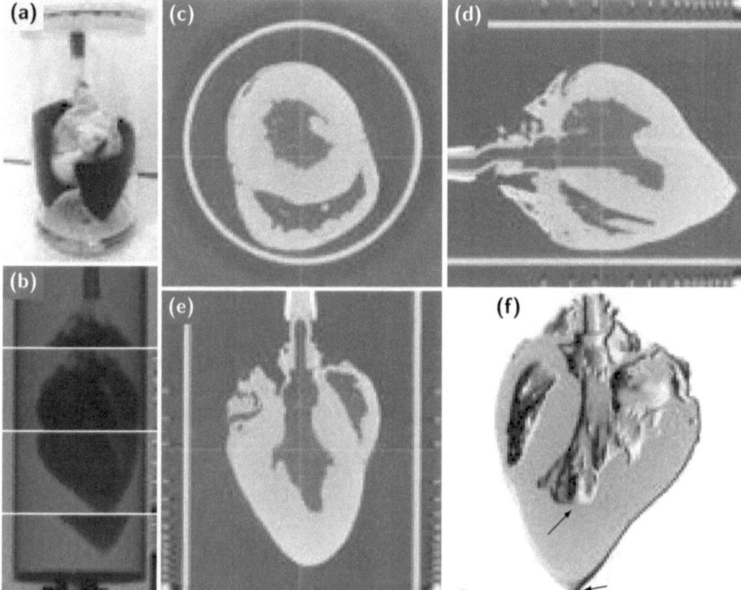

Figure 6.2 – CT scan example of a pig heart. (a) Large sample container with pig heart, stabilized with soft foamed material. (b) Original X-ray images for different vertical positions of rotation stage. Each image is composed of two images, for which the detector had to move back and forth. (c – e) Perpendicular cross sections of final CT reconstructed volume. (f) 3D volume visualization with clipping. Arrows indicate artifacts due to residual liquid inside and outside of the ventricle.

which is almost invisible in the scan. After insertion, the heart was kept inside the container for 60 minutes, to let slight shape deformations due to gravity and water evaporation settle before start of the scan. Due to its size, a pig heart cannot be scanned in one pass with the nanotom CT scanner. Three to four different vertical positions are necessary to cover the whole heart with enough overlap between the individual acquisitions (fig. 6.2b). In addition, the detector screen has to move back and forth horizontally for acquisition of the image of each rotational position, to expand the field of view and capture the entire diameter of the container. Finally, to reduce noise, 8 images (exposure time 250 ms) are taken and averaged for each of the 1800 rotational positions. The entire acquisition takes several hours. Tomographic reconstruction and volume stitching are performed automatically by the software phoenix datos|x. The spatial resolution of the final volume is about 0.0564 mm per voxel.

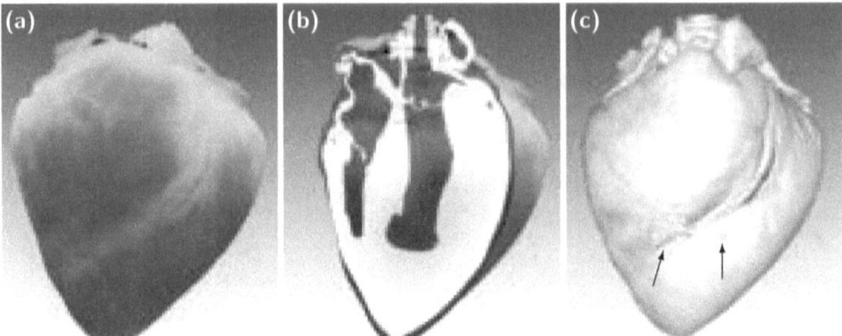

Figure 6.3 – **Comparison of CT volume and surface mesh.** Volume was aligned (only position and rotation, not scaling) to optically reconstructed surface mesh. **(a)** Textured surface mesh. **(b)** Cutaway of mesh with cross section of volume. Note slightly smaller size of volume. **(c)** Iso surface visualization of volume. Arrows indicate residual liquid where foamed material touches the heart.

6.3   Results

Figure 6.3 shows the result of registering the CT scanned volume of a pig heart to the surface mesh reconstructed from camera images. Best-fit alignment of the position and rotation was done manually, no scaling transformation was applied. While the overall shape is very similar, some degree of discrepancy can be recognized. The CT scan has only about 95 % the size of the surface mesh. This is most likely due to the fact that the heart was pressurized by perfusion during acquisition of the photographs for shape reconstruction, while the CT scan shows the preserved heart in a non-pressurized state.

6.4   Discussion

The methods presented in this short chapter are neither novel nor inventive. Nonetheless, they represent a valuable contribution to the whole experiment for future applications, as the internal anatomy can be studied at very high spatial resolution. Depending on the research question, respective regions of interest could be highlighted using contrast agents. For example, the vascular tree can be enhanced by injecting an aqueous suspension of barium sulfate, $BaSO_4$. Alternatively, iodide can be used in an aqueous solution of I_2 and KI. Hornung [80] discusses the respective (dis)advantages of these methods. Some diseases such as atherosclerosis can be visualized right away without contrast agents. Atherosclerotic cardiovascular disease is a hardening and narrowing of the heart's arteries. The calcium contained in accumulated plaques is easily visible in a CT scan [55]. The identification of underlying anatomical causes for abnormal observations in the dynamic excitation propagation is key for the study of arrhythmia. The CT volume data can also be used to create a 3D electrophysiological computer model, for simulation of excitation dynamics and comparison with the multimodal experiment.

For better conservation of the heart shape, it may be advantageous to perform the preservation not only by passive immersion in the fixation solution, but instead by active retrograde perfusion under the same pressure as during the experiment. Using the Langendorff perfusion setup itself for this task has to be ruled out, for reasons of practicality and formaldehyde's toxicity. Instead an encapsulated perfusion device would be useful, that can recycle the excess fixative solution and can be placed inside or connected to a fume hood. Laflamme et al. [109] used such a device for perfusion-fixation of hearts at iso-arterial pressure, and report that post-mortem histologic evaluation of perfused coronary artery shows stenosis comparable to angiography, whereas they confirmed a significant discrepancy between the amount of stenosis in passively fixed coronary arteries and coronary angiography, as previously reported.

Future work will focus on the automatic registration of volume data to the surface mesh, as well as on the deformation according to 3D motion tracking data obtained with cameras (chapter 7) and 4D ultrasound (chapter 8).

For selected experiments, it may be beneficial to use another CT scanning facility for reconstruction of muscle fiber orientation, e.g. using synchroton or laboratory phase-contrast micro-CT [89, 111].

Chapter 7

3D surface deformation tracking

7.1 Introduction

The 3D panoramic optical mapping method presented in the previous chapters has a major limitation, which is the prerequisite of complete elimination of sample movement, in order to suppress motion artifacts (cf. section 2.1.3). These artifacts can be as strong as to completely obscure the fluorescent signals of the electrophysiologic excitation dynamics. Therefore, the heart has to be immobilized by administration of chemical agents which inhibit the contractile function of the cardiomyocytes. However, as the main function of the heart lies in its acting as a pump for blood, normal and dysfunctional contraction of the heart muscle is an important quantity of measurement in basic cardiac research. Furthermore, while modern excitation-contraction decoupling agents like myosin II inhibitor blebbistatin leave the electrical dynamics mostly unaffected, their toxicity and known [71] as well as possibly unknown side effects cannot be neglected. It is therefore desirable to be able to perform 3D panoramic optical mapping on an uninhibited contracting heart, in order to study the electrical and mechanical dynamics and their interplay at once.

7.1.1   2D and 3D motion tracking methods

In the case of individual camera videos, 2D motion tracking of natural tissue features and co-moving analysis of the optical signals *is* possible, as has been demonstrated by my colleague Jan Christoph [82]. The method is based solely on computational post-processing of the fluorescent optical mapping videos and does not require special lighting, fiducial markers, additional hardware or other auxiliary means. Locally enhancing the image contrast boosts the structure of natural tissue features and allows to register the movement of the heart surface using established motion tracking techniques. Unfortunately, this method is difficult to extend to 3D motion tracking of large parts of the heart.

Traditional, passive stereoscopic methods for 3D geometry reconstruction and motion tracking require that the 2D locations of surface feature points can be reliably identified in both images of a pair of calibrated cameras. However, this is only possible if the images are not too different, limiting the spatial separation of the cameras. For large baselines, natural features on the surface of the object to be tracked appear too dissimilar, and stereo correspondences cannot be established. Using this technique for 360° motion tracking of a heart would therefore require a very large number of cameras.

To circumvent this problem of dissimilar natural features under large viewing baselines, artificial markers can be attached to the sample, whose 2D positions can be reliably identified under different viewing conditions. In a study by Bourgeois et al. [60], ring-shaped markers were attached to the epicardial heart surface, in order to perform 3D motion tracking of and optical mapping from the sites encircled by the markers. This method was later extended to use dot-shaped markers [87], which allows to acquire membrane potential signals "from all sites within the marked region except those obscured by markers."[1] Motion tracking of these sites within a marker triangle was done by linear interpolation, assuming "that the triangle is small enough that deformation is homogeneous within the triangle,"[2] which they considered granted when placing markers at 8 mm spacing on a pig heart. Both studies applied excitation ratiometry to reduce motion artifacts caused by inhomogeneous illumination.

Instead of attaching real markers, active stereo vision techniques create virtual features on the heart surface. Laughner et al. [68] and Wang et al. [77] demonstrated the use of structured light imaging for mapping of cardiac surface mechanics. This technique can be compared to passive stereo vision, where one camera is replaced by a DLP (Digital Light Processing) light crafter, which projects a series of repeating binary fringe patterns into space. These stripe patterns appear bent when they hit onto the surface of objects and are observed by the camera from a different viewing location. This allows to reconstruct the 3D geometry of a deforming object at high spatial and temporal resolution. While the structured light imaging technique can capture the 3D shape of an object for each time step, it cannot register motion parallel to the surface. To form correspondences between consecutive 3D frames and derive deformation data, the authors employed a nonrigid motion tracking algorithm, that registers and warps a template geometric shape of one scan to each other time step. The strength of this method is that it does not rely on intrinsic surface features or fiducial markers. However, the method was shown with one camera only, and only provided mechanical tracking data. The authors described it as first step towards whole heart mapping. While there are methods for multi-camera multi-projector structured light scanning being developed [84], to my knowledge an application on a whole beating heart with simultaneous optical mapping has not yet been demonstrated. Furthermore, the presence of a perfusion bath adds another source of complexity to structured light imaging.

In our quest for full 360° panoramic optical mapping and motion tracking system, we would like to have a method that does not need additional equipment such as

---

[1] Quote from Zhang et al. [87, p. 440].
[2] Quote from Zhang et al. [87, p. 441].

light crafters, as the experimentation table is already very crowded by cameras, lights and other appliances like holders for stimulation electrodes. Encouraged by the good 2D motion tracking results of my colleague [82], hope was given to establish a new purely image-based, passive 3D motion tracking method by combination of 3D static shape reconstruction and multi-camera 2D motion tracking of texture features without markers.

7.1.2 Idea for new 3D motion tracking method

For our application requirements we do not need to to have an all-purpose method that can reconstruct arbitrary 3D shapes for each time scan. The following aspects allow for simplification of the problem. While the heart surface does deform during contraction, its general shape does not change dramatically or unexpectedly in the experiment. Furthermore, the heart's movement is restricted by the experimental setup. It may swing and rotate about up to $\pm 20°$, but stays in the field of view of the cameras. We can reconstruct the shape and texture of the heart surface at high spatial resolution, if the heart is immobilized by contraction inhibiting agents. By reconstructing such a static reference mesh at the end of an experiment, it should be possible to deform the mesh in such a way, that it matches the 2D displacement data of all cameras, acquired from previous videos of the contracting heart by 2D motion tracking each frame with respect to the reference time.

From this idea, my colleage Jan Christoph and myself developed the new tracking method and successfully demonstrated it in proof-of-concept experiments on isolated hearts of New Zealand White rabbits ($N = 2$). Additionally, a novel approach for reduction of motion artifacts based on light field estimation was presented. Methods, experiments and results were published in [88]. The main methodological concepts and results will be described in the following. Since the animal species is irrelevant for the proposed method, details of animal handling and specific drug concentrations etc. will be omitted. Notation of symbols in formulae and wording was adapted to be more concise and consistent with the other chapters of this book. Some aspects were clarified where it seemed necessary. The contents of this chapter should be considered a citation of the original publication [88] without using quotation marks everywhere. The combination of 3D surface deformation tracking and 3D panoramic optical mapping with lightfield correction will be referred to as 3D electromechanical optical mapping (3D-EMOM).

7.2 Experimental setup and protocol

Objective of the experiments was to acquire fluorescence videos of a *contracting* heart in sinus rhythm (SR), followed by 3D shape reconstruction of the heart in a *motionless* reference state. The experiments were performed on the panoramic optical mapping setup (compare chapter 2). The four high-speed EMCCD (electron-multiplying charge-coupled device) cameras (fluorescence cameras) were rearranged

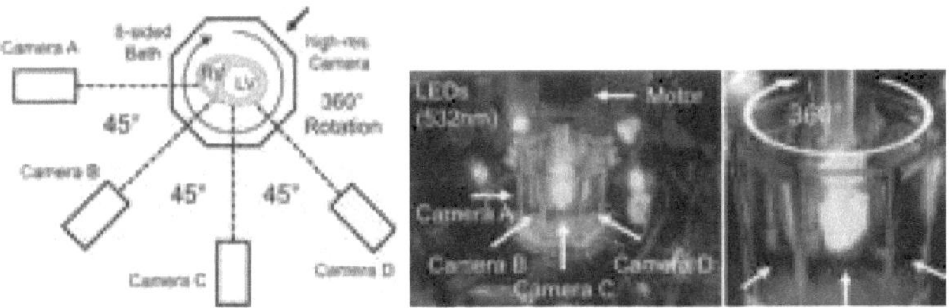

Figure 7.1 – Arrangement of cameras for 3D surface deformation tracking.
Cameras are arranged with 45° separation, looking at half of the hearts surface.
— Images reproduced from [88] with authors' permission.

to look at one half of the heart with 45° separation (see fig. 7.1). The high-resolution DSLR (digital single-lens reflex) camera (geometry camera) was positioned at the back. Experiments were conducted as follows:

1. Alignment of all cameras and acquisition of calibration pattern images.
2. Animal anaesthetization, extraction of heart and connection to perfusion system.
3. Loading of heart with potentiometric fluorescent dye di-4-ANEPPS. Wait until fluorescence strength at maximum.
4. Recording of optical mapping videos of the *contracting* heart during SR with the synchronized fluorescence cameras.
5. Administration of Blebbistatin. Wait until contraction completely inhibited.
6. Rotation and image acquisition of *static* heart for shape reconstruction.

The last step included documentation of the heart shape using the geometry camera under white light, and additional documentation of the surface texture using the fluorescence cameras under illumination with green excitation LEDs.

Camera calibration and static 3D heart shape reconstruction was performed without modification as explained previously in chapters 3 and 4. The time span of these proof-of-concept experiments was kept short, in order to minimize change of the hearts' appearance due to bleaching, internalization, redistribution, or washing out of the dye in the tissue. Overall duration of steps 4 to 6 was approximately 20 min.

7.3   2D motion tracking

The method for 2D motion tracking of cardiac tissue in optical mapping videos presented here was developed by my colleague Jan Christoph [82, 96]. It uses contrast enhancement and a standard optical flow algorithm to track the movement based on the unique texture of the cardiac tissue as it appears in fluorescent optical mapping. This spatial variation of fluorescence intensity is a product of natural tissue inhomogeneities (like blood vessels or fatty tissue) and an non-uniform distribution of dye, see top row of fig. 7.2.

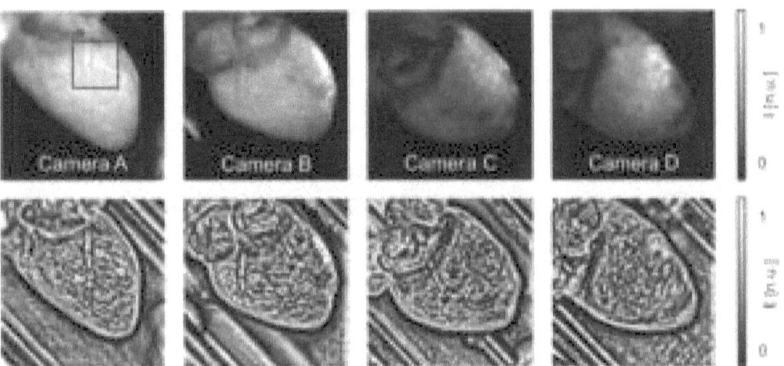

Figure 7.2 – Contrast enhancement of optical mapping videos. Top: original images I^k for each camera $k \in [A, B, C, D]$. Bottom: contrast-enhanced images E^k.
— Images reproduced from [88] with authors' permission.

The fluorescence intensity does further vary in time, due to the purpose of the dye to visualize dynamic processes in the tissue, such as membrane potential or intracellular calcium concentration. Therefore, excitation waves can be observed moving over the tissue, as discussed in section 2.1.2. We want to track the movement of the tissue itself, without the algorithm being distracted by the apparent movement of these excitation waves inside the tissue. For this reason, the videos of each camera $k \in [A, B, C, D]$ were pre-processed with a local contrast enhancement method, in order to emphasize the tissue's small spatial features and suppress the longer gradients of the dynamic wave phenomena, as well as variation in lighting. Each pixel's intensity value $I^k(x, y, t)$ of a video frame $t \in [1, ..., N]$ was renormalized with respect to the minimal and maximal intensities on a circular neighborhood around the pixel's location (x, y):

$$E^k(x, y, t) = \frac{I^k(x, y, t) - \min(\mathcal{S}^k(x, y, t))}{\max(\mathcal{S}^k(x, y, t)) - \min(\mathcal{S}^k(x, y, t))} \tag{7.1}$$

where

$$\mathcal{S}^k(x, y, t) = \{I^k(x', y', t) \mid (x' - x)^2 + (y' - y)^2 < r^2\} \tag{7.2}$$

with a typical diameter $d = 2r$ of 5 to 7 pixels. Resulting video frames with pronounced unique feature pattern are shown in the bottom row of fig. 7.2.

Two-dimensional optical flow was then computed on the pre-processed videos using a Lucas-Kanade image registration algorithm [28] as follows. Subscript symbol c will refer to a video of the contracting heart, whereas s denotes an image of the static, motion-inhibited heart, taken at the end of the experiment after administration of Blebbistatin. These reference images taken with the fluorescence cameras show the heart in the same shape as the images taken with the geometry camera,

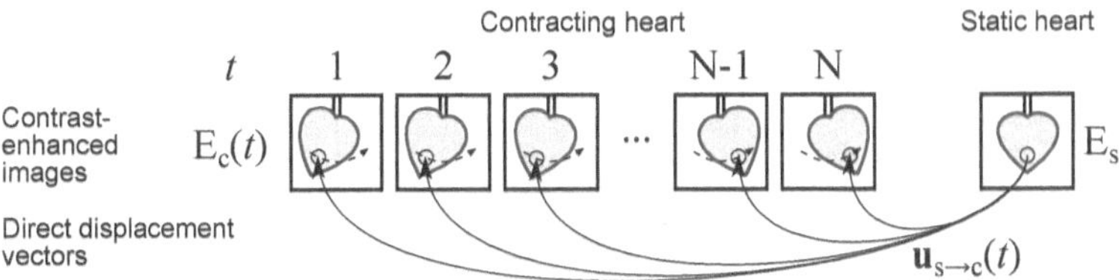

Figure 7.3 – Direct 2D motion tracking scheme between individual contrast-enhanced video frames $E_c(t)$, $t \in [1, ..., N]$ of the contracting heart and a reference image E_s of the static heart, obtained at the end of the experiment after administration of Blebbistatin. Direct displacement vectors $\mathbf{u}_{s\to c}(t)$ are computed between the reference image and each video frame. Note: Camera index k and pixel coordinate (x, y) omitted. — Image adapted from [88].

from which the high-resolution 3D mesh was reconstructed. 2D displacement vectors $\mathbf{u}_{s\to c}^k(x, y, t) \in \mathbb{R}^2$ were determined for each pixel (x, y) of every frame t. The subscript $s{\to}c$ illustrates, that the displacements were computed with respect to the contrast-enhanced image E_s^k of the static heart, and *not between consecutive* video frames of the contracting heart. Thus, the optical flow describes the 2D apparent deformation of the static image to an individual video frame, see fig. 7.3.

7.4 Alternative 2D tracking scheme

Another, alternative 2D tracking scheme was developed, for the case that there were difficulties in the direct computation of the optical flow for some frames, because of too large dissimilarities of the images E_s^k and $E_c^k(t)$, whereas the intra-sequence optical flow of the video could be computed without problems. The dissimilarities could arise from strong swinging motions of the heart during SR. The following two-step tracking scheme was devised. In the sequence of video frames, a reference image $E_r^k := E_c^k(t{=}r)$ with $r \in [1, ..., N]$ was manually selected to depict the contracting heart in a similar form and position as the static heart (cf. fig. 7.4). For sinus rhythm, this usually is the case shortly before a contraction-initiating excitation of the atria and ventricles. The optical flow $\mathbf{u}_{r\to c}^k(x, y, t)$ was computed for each video frame with respect to the intra-sequence reference image E_r^k. Together with the displacements $\mathbf{u}_{s\to r}^k(x, y)$ from the static reference to the intra-sequence reference, the direct displacement field can be expressed as:

$$\mathbf{u}_{s\to c}^k(x, y, t) = \mathbf{u}_{s\to r}^k(x, y) + \mathbf{u}_{r\to c}^k(x, y, t) \tag{7.3}$$

In order to further suppress computational artifacts, $\mathbf{u}_{s\to r}^k(x, y)$ was approximated from the whole stabilized video sequence as follows. First video frames were warped to produce images $E_w^k(t)$, which all look similar to the reference frame. Then, displacement vectors $\mathbf{u}_{s\to w}^k(x, y, t)$ were computed between the static image and each warped video frame. Where this was not possible, the displacement data of the corresponding pixel and time was set to be invalid. An averaged displacement

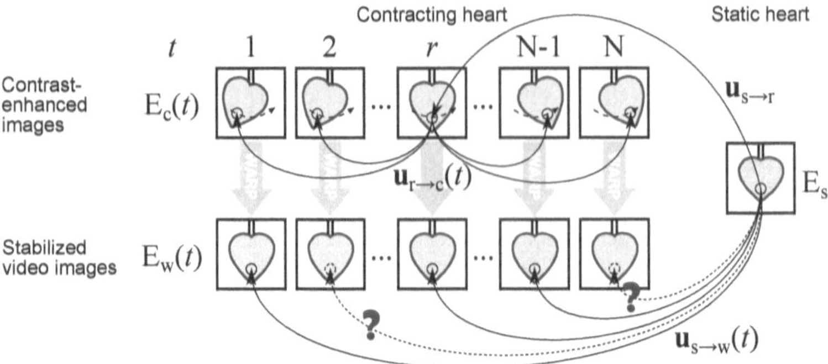

Figure 7.4 – Alternative 2D motion tracking scheme. Optical flow between static image E_s and each frame $E_c(t)$ of the contracting heart video is computed in two steps. First, displacements $\mathbf{u}_{r \to c}(t)$ are computed with respect to an intra-sequence reference image at time $t = r$ and the image sequence warped accordingly. Second, displacements $\mathbf{u}_{s \to w}(t)$ are computed between the static image and warped video frames, where possible (question marks indicate impossibility). From this an averaged displacement field $\bar{\mathbf{u}}_{s \to w}$ is computed to approximate direct displacements $\mathbf{u}_{s \to c}(t) \approx \bar{\mathbf{u}}_{s \to w} + \mathbf{u}_{r \to c}(t)$. Note: Camera index k and pixel coordinate (x, y) omitted. — Image adapted from [88].

$\bar{\mathbf{u}}_{s \to w}^k(x, y)$ was computed for each pixel individually, from all $N^* \leq N$ times t^* for which the previous operation did succeed:

$$\bar{\mathbf{u}}_{s \to w}^k(x, y) = \frac{1}{N^*} \sum_{t^*} \mathbf{u}_{s \to w}^k(x, y, t^*)$$

Finally, the direct optical flow could be approximated:

$$\mathbf{u}_{s \to c}^k(x, y, t) = \mathbf{u}_{s \to r}^k(x, y) + \mathbf{u}_{r \to c}^k(x, y, t) \tag{7.4}$$

$$\approx \bar{\mathbf{u}}_{s \to w}^k(x, y) + \mathbf{u}_{r \to c}^k(x, y, t) \tag{7.5}$$

7.5 3D surface deformation tracking

In this section, vectors in three-dimensional world coordinates will be denoted by upper-case characters, whereas lower-case symbols signify vectors in two-dimensional image coordinates. Furthermore, let $\mathbf{pr}^k : \mathbb{R}^3 \to \mathbb{R}^2$ be the vector-valued function that projects a point from world to image coordinates, according to model eq. (3.3) with calibrated parameter set of camera k.

The three-dimensional movement of the epicardial surface of the contracting heart can be described as a deformation of the static 3D mesh. In this case, the trajectory of an individual mesh vertex i can be written as the sum of the static position vector and a time-dependent displacement vector:

$$\mathbf{X}_i(t) = \mathbf{X}_i^s + \mathbf{U}_i(t) \tag{7.6}$$

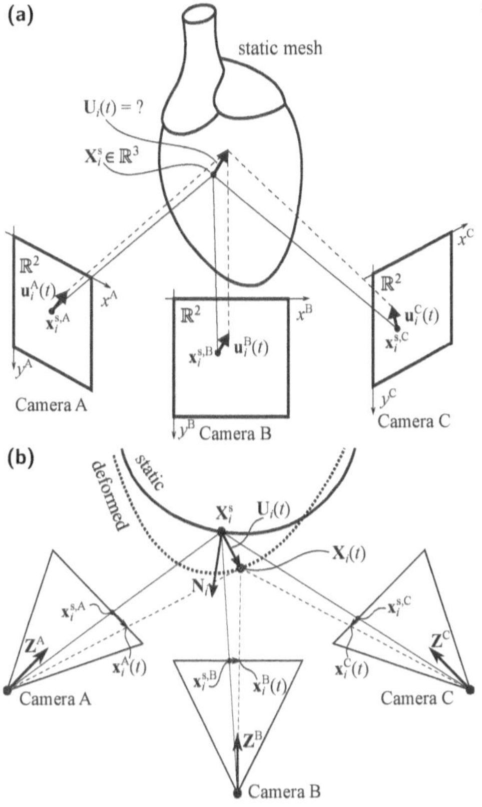

Figure 7.5 – Estimation of 3D surface deformation from 2D displacement data of multiple camera views. Schematic drawings (not to scale) showing the reconstruction of the trajectory for a single vertex i of the triangular mesh. **(a)** Front view, illustrating the projection of the vertex's static position $\mathbf{X}_i^s \in \mathbb{R}^3$ to image coordinates $\mathbf{x}_i^{s,k} \in \mathbb{R}^2$ for all cameras $k \in [A, B, C]$. Evaluating the displacement fields at these locations for time t yields displacement vectors $\mathbf{u}_i^k(t)$. **(b)** Top view, illustrating the estimation of the vertex's position $\mathbf{X}_i(t) \in \mathbb{R}^3$ on the deformed heart surface (dotted curve), by employing an automatic optimization procedure which minimizes the deviations of its projections from the target locations $\mathbf{x}_i^k(t) \in \mathbb{R}^2$ (symbolized by dashed lines). Angle between surface normal $\mathbf{N}_i$ and optical axis $-\mathbf{Z}^k$ (both $\in \mathbb{R}^3$) determines weight of respective camera in objective function, eq. (7.10).
— Images adapted from [88].

However, the true three-dimensional displacement $\mathbf{U}_i(t)$ is unknown. Only two-dimensional displacement data $\mathbf{u}_{s\to c}^k(\mathbf{x}, t)$ in the views of the cameras are known, from which we have to derive the trajectory as good as possible. Figure 7.5 shows a schematic of this situation for a specific vertex i and time t. We start by projecting the position vector $\mathbf{X}_i^s$ of the static mesh to image coordinates of all cameras:

$$\mathbf{x}_i^{s,k} = \mathbf{pr}^k(\mathbf{X}_i^s) \qquad \forall k \tag{7.7}$$

The 2D displacement fields of the cameras are evaluated at these locations for given time t:

$$\mathbf{u}_i^k(t) = \mathbf{u}_{s\to c}^k(\mathbf{x}_i^{s,k},\, t) \qquad \forall\, k \tag{7.8}$$

Here, sub-pixel resolution is achieved using interpolation of the displacement field. Then the displaced *target position* of the vertex in the camera images can be computed:

$$\mathbf{x}_i^k(t) = \mathbf{x}_i^{s,k} + \mathbf{u}_i^k(t) \qquad \forall\, k \tag{7.9}$$

One is tempted to call this quantity $\mathbf{x}_i^k(t)$ the *trajectory of the vertex in image coordinates*, however this is only true for those cameras, for which this point on the heart surface is actually directly visible, and not facing away from the camera.

We cannot simply compute the true 3D position $\mathbf{X}_i(t)$ of the displaced vertex in world coordinates by back-projection of rays and calculation of the intersection point, due to unavoidable uncertainties. Therefore we estimate it using a Nelder-Mead-optimization process [2]. This standard general-purpose optimization routine is initialized with the vertex's position of the static mesh, which will be varied until optimal position is found. Best match with the 2D displacement data is achieved, if the global minimum of following objective function is found:

$$f\left(\mathbf{X}_i(t)\right) = \sum_k w_i^k \cdot \left\|\mathbf{x}_i^k(t) - \mathbf{pr}^k\left(\mathbf{X}_i(t)\right)\right\| \tag{7.10}$$

The objective function computes a weighted sum of the quadratic differences between the previously computed 2D target locations and the 2D projections of the current 3D estimate $\mathbf{X}_i(t)$ for all cameras. The weights w_i^k ensure, that the contribution of a camera is correctly balanced according to its viewing angle onto the surface point, or completely suppressed if facing away. For this, we compute the angle $\varphi = \angle(\mathbf{N}_i, -\mathbf{Z}^k)$ in 3D space between the surface normal and the optical axis[3] of the respective camera (see fig. 7.5b) and define the weights to be:

$$w_i^k = \begin{cases} \cos\varphi & \text{if } \varphi < \hat{\varphi}, \\ 0 & \text{otherwise.} \end{cases} \tag{7.11}$$

The cutoff-angle $\hat{\varphi} = 80°$ was chosen to be less than $90°$, in order to suppress border effects near the edge of the heart, where computation of the optical flow was affected by the static background. Furthermore, the weights have to be non-zero for at least two camera views, otherwise the position cannot be estimated for that vertex.

The size of the surface area for which this 3D motion tracking can be applied, depends on the number and overlap of camera views. With our 4 cameras separated by $45°$ roughly half of the heart can be tracked. The result is a deforming 3D mesh, which closely follows the motion of the real heart surface. The deforming mesh data can be readily used to project the raw fluorescent camera videos of the contracting heart onto it for each corresponding time step, as described in chapter 5.

7.6 Light field estimation and correction

Three-dimensional motion tracking techniques allow for quantification of the deformation of the heart surface, and in combination with 3D panoramic optical mapping a co-moving analysis eliminates motion artifacts, which are caused by the movement of the tissue relative to the cameras. However, motion artifacts caused by relative motion between tissue and light sources remain. This relative motion is due to the

---

[3]The negative sign of $\mathbf{Z}^k$ in computation of φ is due to the cameras looking down the z-axis.

inhomogeneity of the light field produced by multiple light sources. Movement of the heart surface in an ideal, homogeneous light field would not cause such motion artifacts, as there would be no effective relative motion. Experimentally, the preparation of a homogeneous illumination from all sides would be extremely challenging.

To address this problem, previous works applied excitation ratiometry [60, 87], where illumination uses two colors which are alternating switched on and off for odd and even frames. The principle of this method is that voltage sensitivity is different for the two excitation wavelengths, while other sources of fluorescence fluctuations such as motion are common, and are canceled out by division. However, this technique also requires the ratio of both excitation colors to be spatially uniform, which requires special mixing optics and can be difficult to establish for a large panoramic optical mapping setup.

During my work on the 3D motion tracking method, I realized that there is another solution that addresses problem at the heart. While ratiometric imaging eliminates the effects of an otherwise unknown inhomogeneous light field, it should be possible to estimate the spatial variation of the light field itself — without additional aids! Everything needed is already provided by the 3D motion tracking and panoramic optical mapping: The heart surface itself can serve as a probe, which 'measures' the strength of illumination as it moves through space. The combination of a dense set of reconstructed 3D trajectories $\mathbf{X}_i(t)$ for each vertex i of the mesh, and fluorescence intensities $I_i(t)$ recorded along these paths, considering only times where the heart is at rest, allows us to estimate the local strength of the light field, and eventually correct the motion artifacts induced by its spatial inhomogeneity.

A simple model of the recorded fluorescence intensity comprises three factors

$$I_i(t) = l_i(X, Y, Z, t, ...) \cdot f_i(t) \cdot V_i(t) \tag{7.12}$$

which will be described and further simplified in the following. Here $I_i(t)$ denotes the fluorescence intensity taken from the combined texture video at texture coordinates of the i-th vertex, not be confused with the intensity $I^k(x, y, t)$ of individual camera k.

First, $l_i(X, Y, Z, t, ...)$ is the strength of illumination received by the surface element at vertex i as it moves through space, which depends on location, time, and other unknown parameters such as position and direction of light sources. Assuming that all of this complexity can be approximated by a time-constant light field $L(X, Y, Z)$, we get:

$$l_i(X, Y, Z, t, ...) \approx L\big(\mathbf{X}_i(t)\big) \tag{7.13}$$

Second, $f_i(t)$ is the strength of emitted fluorescent light, that depends on the local concentration of dye. This can be considered to be constant over the course of a typical recording:

$$f_i(t) \approx f_i \tag{7.14}$$

Third, $V_i(t)$ describes the time-dependent modulation of fluorescence proportional to the membrane potential due to the potentiometric nature of the dye. We take advantage of the fact that the membrane potential is at resting potential during the diastolic interval (DI) of sinus rhythm:

$$V_i(t) \approx V_i^{\mathrm{rest}} \qquad \text{for } t \in \mathrm{DI} \tag{7.15}$$

Inserting these approximations into eq. (7.12) and reordering yields:

$$L\big(\mathbf{X}_i(t)\big) \approx \frac{I_i(t)}{f_i \cdot V_i^{\mathrm{rest}}} \qquad \text{for } t \in \mathrm{DI} \tag{7.16}$$

The value of $f_i \cdot V_i^{\mathrm{rest}}$ is proportional to the resting (or mean) fluorescence intensity at vertex i, and can be obtained from the corresponding coordinate in the fluorescence texture image, that is created for the static heart mesh (see fig. 7.6c). By inserting all the measured fluorescence intensities $I_i(t)$ of all vertices at times during diastole, the right-hand side of eq. (7.16) gives a large data set of values for the strength of the light field at positions along the corresponding trajectories $\mathbf{X}_i(t)$.

An estimate $L^{\mathrm{est}}(X, Y, Z)$ of the light field needs to be generated from these data using suitable modeling methods, which are capable of smoothed interpolation as well as extrapolation, since the data are noisy and possibly sparse. This allows to evaluate the estimated light field along a trajectory for all times, i.e. not only during diastole but also during systole. Then fluorescence time series can be corrected for illumination-induced motion artifacts:

$$I_i^{\mathrm{corr}}(t) = \frac{I_i(t)}{L^{\mathrm{est}}\big(\mathbf{X}_i(t)\big)} \qquad \text{for all } t \tag{7.17}$$

As the heart is illuminated by several LED spotlights, it would be difficult to construct a mathematical model of the light field, that recreates the complex physical arrangement of the light sources (position, orientation, beam profiles, ...) and can be globally fitted to the experimental data of eq. (7.16). For the proof-of-concept we employed a very simplistic approach and estimated the light field at position $\mathbf{X}_i(t)$, by locally fitting a three dimensional polynomial P_n using all data during diastole from neighboring trajectories within a small sphere with radius $r = 2\,\mathrm{mm}$.

$$L^{\mathrm{est}} = P_n(X, Y, Z) = \sum_{m=0}^{n} \alpha_m X^a Y^b Z^c \qquad a + b + c \le m \tag{7.18}$$

Polynomials of up to third order ($n = 3$) have been tried for the proof-of-principle experiments, and we found that for these data linear ($n = 1$) approximation gave best results.

7.7 Experimental results

This section presents some of the results of the proof-of-principle experiments published in [88]. Figure 7.6 shows the result of the 3D shape reconstruction of rabbit

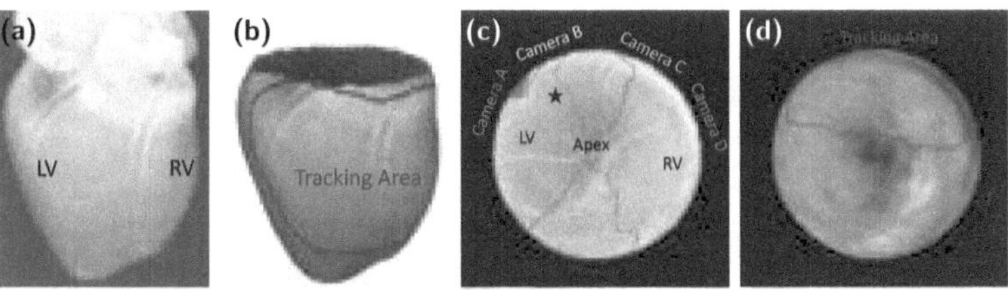

Figure 7.6 – Results of static 3D surface reconstruction. (a) Photo of heart in perfusion bath taken with high-resolution geometry camera. (b) Rendering of mesh with color texture applied. Area for 3D motion tracking with four cameras (blue outline) covers almost 180° of the heart's circumference and most of the ventricles' height. (c) Polar texture map of fluorescence on heart surface, also indicating the tracking area. (d) Polar map of the heart surface, showing the overlapping regions visible to the individual cameras. — Images reproduced from [88] with authors' permission.

heart #2, generated from 72 images obtained in the last step of the experimental protocol (see section 7.2). A high-resolution triangular mesh of the ventricular surface was reconstructed. From this, a smaller sub-mesh was extracted, which was used for application of the 3D motion tracking method. The size of this tracking area depends on the number of cameras and their overlap. For this experiment four cameras arranged at 45° separation were used. 3D tracking can only be performed for vertices, which are visible for at least two cameras (cf. fig. 7.6d). The size of tracking area was further manually reduced, to exclude parts at the boundary where the method obviously produced wrong results, such as runaway vertices. This was necessary, because the method currently does not incorporate any form of regularization. Nevertheless, the final tracking area spans almost 180° of the heart's circumference and most of the ventricles' height from apex to base (see figs. 7.6b and 7.6c).

3D-EMOM results of rabbit heart #2 are shown in fig. 7.7. At the time of video acquisition with the fluorescence cameras, the heart was beating with a regular SR of about 2.5 Hz, which led to a very periodic motion pattern over long times of many cardiac cycles. The motion is composed of the contraction itself and a synchronous swinging of the whole heart. Figure 7.7b visualizes the result of the 3D Surface Deformation Tracking of the contracting heart by plotting the reconstructed trajectories of several vertices on the tracking area. The reconstructed trajectories describe a smooth path with high spatio-temporal resolution and no visible jitter. A single trajectory is depicted in fig. 7.7c and plotted for a complete cardiac cycle. It has an elongated, slightly twisted shape, and closes in on itself almost perfectly due to the periodic activation and resulting motion. The principal axis of the trajectories corresponds to the swinging motion and is aligned perpendicular to the heart surface for most of the tracking area. The color code represents the computed velocity along the paths.

Figure 7.7d visualizes the measured values of the right-hand side of eq. (7.16) along the trajectory. If only the part during diastole is considered (bottom), the

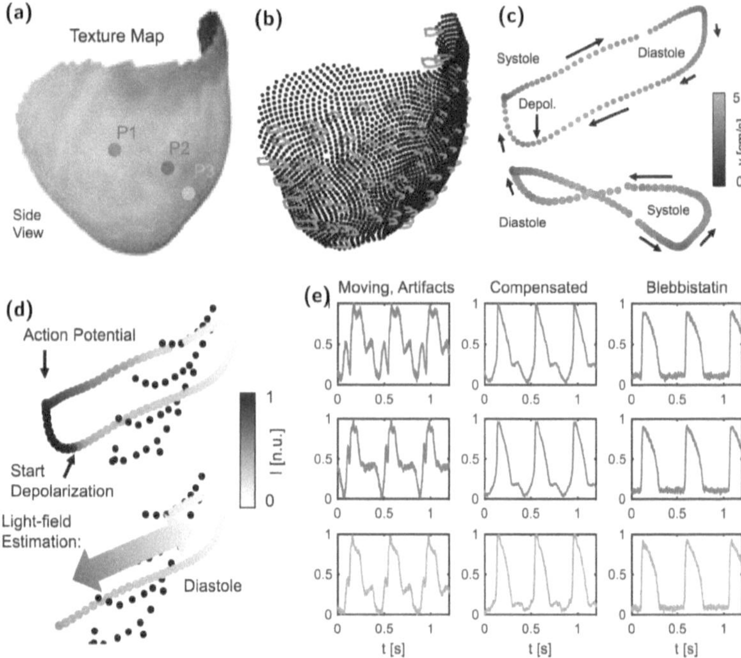

Figure 7.7 – Results of 3D electromechanical optical mapping. (a) Rendered side view of tracking area with fluorescent texture map. **(b)** Trajectories of a selection of vertices obtained by 3D motion tracking. **(c)** Trajectory of a single vertex as seen under different viewing angles. Color code represents velocity along trajectory of a full cardiac cycle. **(d)** Visualization of values of right-hand side of eq. (7.16) along trajectory for times of a full cardiac cycle (top) and only during DI (bottom). **(e)** Time series of three vertices (cf. panel a), showing motion artifacts obtained by 3D panoramic optical mapping of contracting heart without 3D motion tracking (left), the result of optical mapping with motion tracking and light field correction (middle), and action potentials obtained by 3D panoramic optical mapping of the immobilized heart (right).
— Images reproduced from [88] with authors' permission.

variation of fluorescence due to dynamic changes of the membrane potential (action potentials) is eliminated and the spatial gradient of the light field becomes apparent. Neighboring vertices are shown as black dots.

The ability of the 3D-EMOM method to perform a co-moving measurement of the membrane potential for reduction of motion artifacts is shown in fig. 7.7e. Time series of three points on the tracking area (cf. panel a) are shown for three different cases. The left column shows strong motion artifacts, as they arise by performing panoramic optical mapping without surface deformation tracking, i.e. mapping the videos of the contracting heart onto the static mesh. The middle column shows the result of panoramic optical mapping with surface deformation tracking and light field correction applied. The motion artifacts are much less pronounced. For comparison, the right column shows a control recording, obtained by panoramic optical mapping of the motionless heart after administration of Blebbistatin.

7.8 Discussion

The 3D electromechanical optical mapping technique presented in this chapter is capable of tracking the surface deformation of large parts of the heart solely by exploiting the information provided by multiple cameras used for panoramic optical mapping. No fiducial markers attached to the heart surface nor auxiliary external equipment is necessary. Furthermore, the use of the heart surface as a probe for estimation of the spatially inhomogeneous light field facilitates motion artifact reduction without employing ratiometric methods.

The number of data points from which the light field can be estimated depends on the spatial sampling of the mesh and the temporal sampling of the trajectories. The quality of the estimation as presented is furthermore affected by the motion of the heart surface. If the the motion at a specific point on the heart is mostly perpendicular to the surface, then the trajectories of neighboring vertices are more or less parallel to each other and do not overlap much. On the other hand, if the motion is mostly in plane of the heart surface, then trajectories of neighboring vertices are also in this plane and do overlap with each other. Ideally, for good estimation of the light field, we want to have both: overlap of the trajectories *and* broad distribution in three dimensions. Different strategies are conceivable. During ventricular fibrillation (VF), the chaotic uncorrelated excitation patterns may make it possible to use the full trajectories, where contributions from different parts of the action potential (AP) should even out on average. Additionally the irregular motion would provide a better spatial sampling due to many overlaps. Another strategy would involve purposely moving the heart through the bath, e.g. with help of the motorized rotation stage, in order to sample a larger volume of the light field. Yet another approach could use another solid object as a probe before or after the experiment with the heart.

So far we were able to demonstrate the method experimentally using 3 or 4 cameras on two hearts [88]. While this only provided tracking of up to half of the ventricular surface, there are no technical restrictions in increasing the area size by addition of more cameras, for eventually full 360° panoramic tracking. By comparison of the reconstructed trajectories of a vertex using three cameras at 45° versus two cameras at 90° (data not shown here), we anticipated that full 360° motion tracking should be possible using five to six cameras.

Since the 3D tracking is based on 2D tracking which can be applied to each camera separately, neighboring cameras do not have to show similar features and can also be fitted to record different modalities such as membrane potential and intracellular Ca^{2+} concentration using different dyes. Furthermore, the tracking data has the same temporal resolution as the optical mapping videos, which is advantageous to tracking using structured light imaging.

More experimental results and challenges are presented and discussed in our original paper [88]. Much to my own dismay, there was no time yet to do further validation and improvements of the technique. In addition to the outlook and suggentions for improvement mentioned in the discussion section of [88], I propose the following. Validation shall be performed on the basis of computer generated videos, possibly using ray tracing for realistic lighting simulation. This allows to compare the reconstructed lightfield and motion data against the artificial ground truth data. Different heart motion patterns, lighting conditions, and camera arrangements can be analyzed. Technical improvements should involve a break-down of movement into rigid body motion of the whole heart and local deformation. Runaway-vertices can be prevented by introducing regularization with usage of a physical deformation model and restriction of large local deformation with respect to neighboring vertices. Additionally, the local correlation of different camera images projected into the texture map can be incorporated as an additional weight into the optimization of the mesh deformation.

In case future experiments will show that 3D-EMOM has difficulties in tracking of regions with low natural surface features, then these regions could be enhanced by application of fiducial markers, without the need to cover the whole heart. Furthermore, it should be possible to use very small markers, that only diminish the fluorescent light and add structure to the surface, but do not completely block the signal of whole camera pixels.

Chapter 8

4D ultrasound

8.1   Introduction

The optical methods of the previous chapters, namely 3D shape reconstruction, 3D panoramic optical mapping, and 3D motion tracking, all only touch the surface of the heart, so to speak, and cannot reveal the structure, activity or motion *inside* the tissue. One way to reveal these properties is by using ultrasound (US), as the acoustic waves have a deeper penetration depth than light. Ultrasound has a long history with wide-range usage in medical research and diagnostic application, due to its ability to noninvasively generate images of the body's internal structures, such as muscles, tendons, and soft organs. Similarly to radar and sonar, ultrasound can be used to measure the distance to objects by sending out sonic pulses and evaluating the timings and amplitudes of reflected signals. Modern 4D ultrasound probes are capable of scanning three-dimensional volumes in fast succession, allowing observation of beating hearts in real-time. While US cannot measure the electrical dynamics of the cardiomyocytes directly, ongoing research aims to reconstruct the intramural transmembrane action potentials from the induced muscle contractions with data assimilation techniques [100, 110, 113].

The objective of this chapter was to extend the 3D panoramic optical mapping setup with a commercial 4D ultrasound system for simultaneous optical and acoustical acquisition of a beating heart. For integration of the US probe into the experimental setup without obstructing the optical mapping, a new bath with acoustic window was developed. Furthermore, the pose of the probe and geometry of its 3D field of view (FOV) had to be calibrated with respect to the global (optical) coordinate system (CS). A special 3D grid was developed, in order to perform the acoustic calibration in situ within the bath, without the need for an external tracking system.

The developed methods were already partially applied and published in Christoph et al. [95]. Here I will present more methodical details, further development, validation, and discussion within the context of this book.

8.2 Medical ultrasound

8.2.1 Image formation

Ultrasonic waves are generated by a transducer containing a piezoelectric crystal, that expands when a voltage is applied across it. Similarly, a voltage is generated if pressure is applied to the crystal. Therefore the same transducer can be used for emitting and receiving ultrasonic waves (also called transceiver). Traveling through the tissue, the waves are partially absorbed, converting their carried energy into thermal energy (commonly used in high-energy ultrasound therapy for deep heat treatments). At boundaries between media with different *acoustic impedance* $Z = \rho v$ ultrasound waves are partially transmitted and partially reflected. Here ρ is the density of the medium (in $\mathrm{kg\,m^{-3}}$) and v is the speed of sound through the medium (in $\mathrm{m\,s^{-1}}$). The distance to the boundary can be deduced from the time of the received echo, while its amplitude corresponds to the change of impedance. By plotting the amplitude of the echo over time, a profile of the tissue is captured, that reveals the outlines of organs.

8.2.2 Transducer types

While a probe with a single piezoelectric transducer can measure the echos from a beam into the tissue, a linear array transducer (series of transceiver elements in a row) can generate an image of a two-dimensional cross section of tissue, also known as a sonogram. Three-dimensional still images, or volumes, can be acquired by moving a linear array transducer through space in a direction perpendicular to the image plane. On the one hand, this can be a constrained sweeping or rotating motion, which is usually performed by a motorized mechanism. If the movement on the other hand should be unconstrained, it is often performed manually, which is known as *freehand 3D ultrasound*. For both methods, the transformation from 2D image coordinates to 3D space must be calibrated and the pose of the probe tracked in a world frame of reference, in order to reconstruct a geometrically correct volume.

Another faster option for 3D US imaging comes with 2D array transducers. Here, a three-dimensional volume is scanned in one go without moving the US probe. The acoustical FOV can be enlarged over the footprint of the 2D array, by shooting rays into different directions with active beamforming, using modern ultrasound probes with phased array technology. When the acquisition rate is fast enough to produce time-resolved 3D volumetric videos of moving organs such as the heart, such transducers are advertised as *4D ultrasound probes* for live volume imaging (also known as real-time 3D ultrasound), although the transducer array is just a 2D matrix. However, the spatial resolution of such 3D/4D probes typically is lower than that of 2D probes, i.e. linar array transducers.

8.2.3 Freehand 3D Ultrasound – calibration and tracking

The literature on calibration and tracking techniques for ultrasound transducers focuses on freehand 3D ultrasound. Similar to camera calibration (see chapter 3), intrinsic and extrinsic coordinate mappings need to be determined. The extrinsic transformation describes the position and orientation (pose) of the ultrasound probe in a global frame of reference. The intrinsic transformation maps 3D points from local Cartesian transducer coordinates to 2D sonogram image coordinates. Early tracking systems used mechanical positioning of the transducer, where its extrinsic pose can be calculated from the angles of an articulated arm. Further developments used electromagnetic, acoustic and optical tracking methods. The optical techniques seem to be the most suitable for the needs in an operating room, as they do not interfere with other medical equipment and can track multiple tools at the same time. Stereo camera systems were used to triangulate and track the position and orientation of the ultrasound transducer (fig. 8.1). For this, optical markers are attached to the probe, which allow to deduce its pose uniquely. Calibration of the intrinsic transformation is done by scanning the known geometry of a special acoustic calibration object, a so-called *phantom*, which also has optical markers attached to its side. The phantom usually is a liquid-filled box, containing the calibration object. The spatial relationship between the calibration geometry inside the phantom and the optical markers on its outside is know by design. While the probe is moved on the phantom, the sonograms are recorded and the stereo camera system tracks the markers on both objects. This gives the momentary rigid transformation between the markers of the phantom and the probe with respect to a CS defined by the tracking system. The echos of the phantom geometry are located in the sonograms for all recorded times, and a calibration algorithm can finally compute the intrinsic transformation between the 2D ultrasound space and the transducer's local coordinate system, defined by the markers on the probe. Using this information, a 3D volume can be reconstructed, by moving the probe over a patient or subject, and projecting the tracked 2D sonogram images to 3D space, filling the gaps between individual slices with interpolated data. Many different phantoms have been used for calibration, such as spanned nylon threads in a Z shape [30, 78], arbitrary wires [112], and even stairs of Lego® bricks [86]. These objects differ in their capacity for easy, accurate and automated registration of the respective echos in the sonograms. A different approach aims to facilitate 3D freehand US without external tracking using a neural network for image-based estimation of the transducer's motion [102]. For reviews of the different techniques see [39, 40, 92].

8.2.4 3D/4D ultrasound

Mozaffari and Lee state: "Freehand 3-D US systems are more prevalent in the academic environment, whereas in clinical applications and industrial research, most studies have focused on 3-D US transducers and improvement of hardware performance. This topic is still an interesting active area for researchers, and there remain

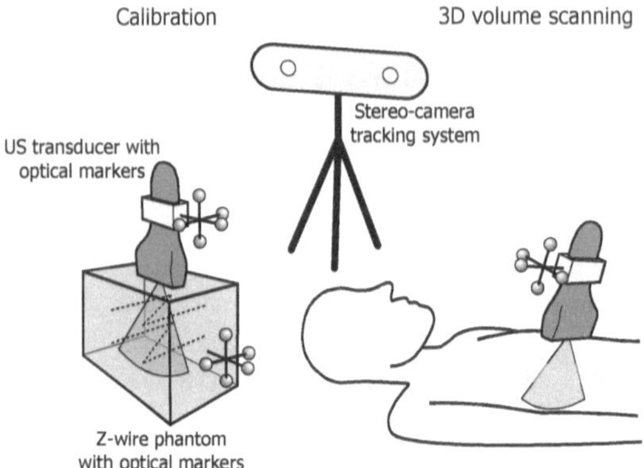

Figure 8.1 – Freehand 3D ultrasound with optical calibration and tracking.
Left: calibration of transducer's intrinsic transformation using stereo-camera tracking
system and acoustic phantom with known internal geometry (Z-wires, dashed lines).
Right: tracking of transducer's extrinsic pose during surgery for 3D reconstruction of
patient's organs.

many unsolved problems to be addressed."[1] 3D/4D probes are capable to scan
volumes in one go without the need to move the transducer. Different calibration
techniques have been proposed, based on an asymmetric angled planes phantom [61],
a tracked needle [85], or an automatic method using 3D printed phantom and an
untracked marker [105]. As commercial medical 3D/4D US systems are intrinsically
calibrated by the manufacturer (for a specific medium density), these calibration
methods aim to estimate an extrinsic rigid transformation between the physical US
probe and its 3D image space, similar to freehand 3D ultrasound. Such extrinsic
calibration is needed, if the position of the transducer with respect to an global coordinate system needs to be known, e.g. for stitching multiple volumes from different
locations together, or to guide surgical interventions. The probe's global pose can
be tracked with optical or mechanical methods as explained above.

In our case we do not need an unconstrained range of motion, as the heart is
not moving much and fits into the acoustic FOV. Therefore the probe's position can
be kept fixed throughout the experiment, and an external tracking system is not
needed. Then the rigid transformation between 3D image space and the physical
probe is irrelevant. We only need to calibrate the transformation between image
space and our global common coordinate system established by camera calibration.
For this a 3D grid phantom was developed, which can be placed inside the perfusion
bath, and allows the calibration to be performed in situ.

---

[1]Quote from Mozaffari and Lee [92, p. 2099].

8.3 Materials and methods

8.3.1 4D ultrasound system

A commercial medical ultrasound system[2] with a fast 4D probe[3] was previously selected and bought by my colleagues (figs. 8.2a and 8.2b). This matrix array transducer has a frequency bandwidth of 1.5 MHz to 3.5 MHz and is designed for transthoracic adult full volume echocardiography. In addition to volume imaging it supports 2D, color Doppler, spectral Doppler, and alternate line phased contrast agent imaging, which are not of interest in the scope of this work. The transducer is well suited for large rabbit and pig hearts, as it facilitates a scanning range up to 300 mm, and elevation and azimuth angles spanning up to 90° (fig. 8.2e). The acquisition rate depends on the settings for size of the 3D field of view and its spatial sampling rates, the latter of which can be set by space/time mode (S1, S2, T1, T2), see table 8.1.

| Volume size | | | Acq. rate | |
El.	Az.	Depth	S2	T2
90°	90°	60 mm	26 Hz	85 Hz
90°	90°	180 mm	11 Hz	36 Hz
90°	90°	300 mm	7 Hz	23 Hz
50°	50°	60 mm	55 Hz	172 Hz
50°	50°	180 mm	23 Hz	73 Hz
50°	50°	300 mm	15 Hz	46 Hz

Table 8.1 – 4D US probe speeds. Excerpt of the possible acquisition rates (volumes per second) of 4D US probe 4Z1c at 2.8 MHz for different volume sizes and spatial resolutions.

Unlike the fluorescence cameras, the acquisition of the ultrasound system cannot be controlled with a TTL (transistor-transistor logic) trigger signal. Also, the acquisition rate of the 4D data is much lower than our typical video frame rate. For post-experiment temporal alignment with video data, the US system was on request customarily equipped with a TTL output, signaling whenever the device is recording. This signal was captured together with the camera triggers using the BIOPAC MP150 data acquisition system. The circular memory buffer of the device can hold up to 10 seconds of volume data. If the acquisition is longer, only the latest 10 seconds of data are saved, and the corresponding start time has to be determined from the end of the TTL output signal. The ultrasound system saves pre-processed[4] 4D volume data and accompanying meta data according to the DICOM (Digital Imaging and Communications in Medicine) standard, commonly used in medical applications.

8.3.2 Perfusion bath with acoustic window

For integration of the ultrasound system into the experiment, I designed a new perfusion bath, which allows for scanning with the US transducer from below. It contains a circular opening in the base, into which a latex membrane can be mounted,

---

[2]ACUSON SC2000 PRIME, Siemens Healthcare.

[3]4Z1c, Siemens Healthcare.

[4]During acquisition, the local contrast at various scanning depths can be adjusted at the US system with sliders. These settings directly affect the displayed and saved data.

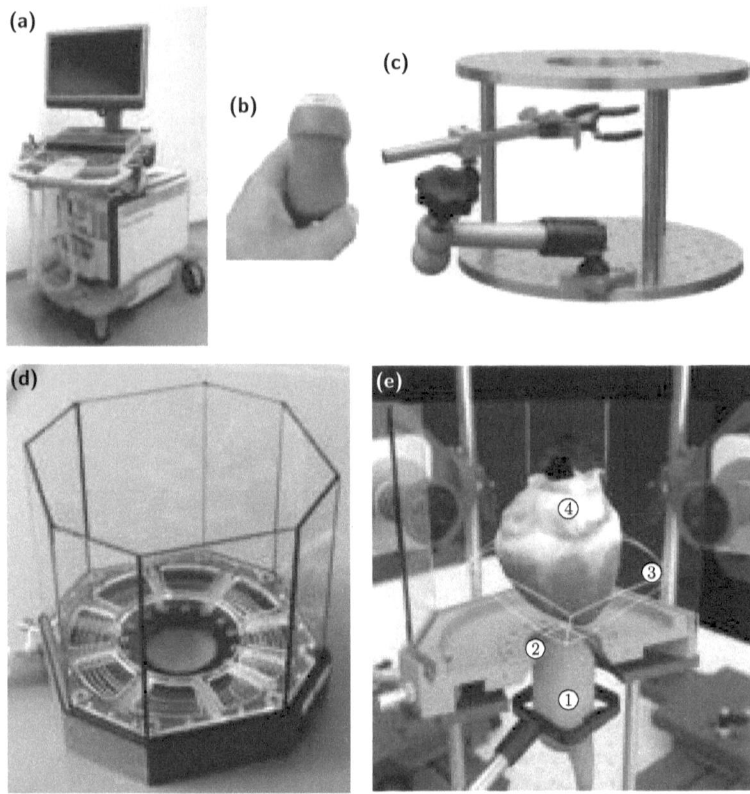

Figure 8.2 – Integration of 4D ultrasound system into the experimental setup.
(a) Ultrasound system ACUSON SC2000 PRIME. (b) 4D-ultrasound probe 4Z1c with
2D phased array transducer. (c) New bath support stand with central opening in top
plate and hydraulic arm with clamp for holding of ultrasound probe. (d) New perfusion bath for large hearts, with integrated acoustic window and accessories in base.
(e) Computer-generated cutaway drawing of bath and stand, showing the ultrasound
probe ① pressing against the latex membrane ② from below. The white wire frame ③
illustrates the acoustical field of view (90° × 80° × 100 mm) of a pig heart ④.

forming an *acoustic window*. While our other baths have an integrated heating in
form of an electrical heating foil below the bottom glass, here heating is realized
with electrical heat wire inside laid around the central opening. The base of the bath
was manufactured with a CNC-machine and integrates aforementioned opening and
heating, as well as a temperature sensor, liquid outflow with two tube connectors,
electrical connectors and grooves for the glass windows. The bath was custom-built
by our workshop and a local aquarium company. The finished bath is depicted in
fig. 8.2d, a cutaway is shown in fig. 8.2e. One tube leads to an external height-
adjustable overflow, the other can be used for quick draining without disconnecting

tubes. The benefit of this integrated design of the base is that viewing through all glass windows is unobstructed.

Additionally, a new bath support stand had to be designed and built, which has a central opening in the top plate and a holding arm for the ultrasound transducer (fig. 8.2c). The stand has a fixed height, but can be rotated on the table around the vertical axis. Threaded holes on a 50 mm raster allow for mounting of additional equipment to the stand. The hydraulic arm facilitates easy 6-DOF (degrees of freedom) adjustment of the transducers field of view and firm fixing with just one knob.

8.3.3 Visualization and resampling of volume data

The 4D data generated by the ultrasound device is stored as a series of 3D volume arrays in a DICOM file. However, the volumes are not parametrized in Cartesian coordinates, but rather in custom spherical coordinates (range, azimuth, elevation) with non-uniform angular spacing of the scan lines, specific to the acoustic FOV geometry of the ultrasound probe. The format is well suited for efficient storage of the data, but the special geometry is not convenient for visualization with simple volume rendering methods, such as Vispy's `VolumeVisual` class which expects a cuboid volume shape. The non-linear equations for conversion between transducer coordinates and local Cartesian coordinates were kindly provided by Siemens, but cannot be disclosed here. Parameters like array size, scanning range, and viewing angles can be read from private tags in the DICOM file for each recording. Using the terminology of camera calibration (see chapter 3), these equations compose the intrinsic transformation, which describes the mapping between local transducer coordinates and corresponding location in computer memory array.

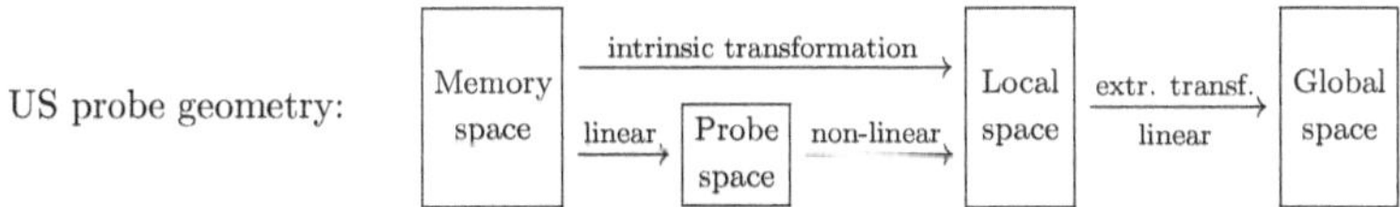

I extended the `VolumeVisual`'s GLSL shader code to support anisotropic data using the provided coordinate transformations. Thus, the 4D volume videos can be directly opened, correctly visualized and played back fast.

Additionally, a set of Python classes were developed, which handle the coordinate transformations of cuboid and pyramid shaped geometries at arbitrary positions and orientations. They can be used to transfer and resample the volume data between different source and target geometries in the overlapping region. The abstract base class `VolumeGeometry` defines methods for transformations between global, local, and array coordinates, the details of which are implemented in the subclasses. Class `VolumeGeometrySiemens4Z1C` describes the special volume geometry of the 4D ultrasound transducer. Class `VolumeGeometryBox` describes a cuboid volume geometry, which is defined by the number of samples in each array dimension, corresponding spatial sampling rates, and a 3-vector of the array origin in local

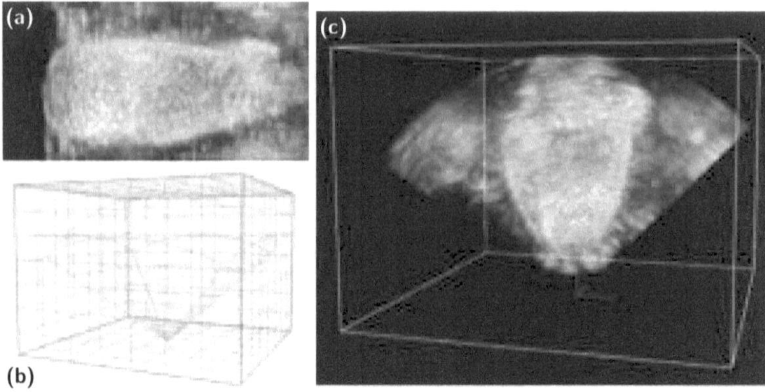

Figure 8.3 – Resampling of volume data. (a) Visualization of original acoustic data of a pig heart (scan-line vs. range projection). (b) Source and target geometries for resampling of volume data from acoustic FOV (green) to Cartesian coordinates of bounding cuboid (red). Both geometries share the same local origin. (c) Volume visualization of acoustic data resampled to box geometry.

coordinates.

Cuboid geometry: $\boxed{\text{Memory space}} \xrightarrow[\text{linear}]{\text{intrinsic transf.}} \boxed{\text{Local space}} \xrightarrow[\text{linear}]{\text{extrinsic transf.}} \boxed{\text{Global space}}$

For both geometries, the linear transformation from global to local Cartesian coordinates is defined by a 4×4 homogeneous affine matrix, which describes the pose relative to the world origin (rotation and translation). Data is resampled by application of the transformations from array coordinates of the source geometry to global coordinates, and then back to array coordinates of the target geometry, using the respective inverse transformations. For the actual resampling operation the function `map_coordinates()` from the Python module `scipy.ndimage` is used with cubic spline interpolation.

A bounding box geometry can be computed from the acoustic field of view (see fig. 8.3b). Different spatial sampling rates can be used for fast preview or final high-resolution resampling. The definition of the box geometry with variable array origin allows it to share the same local origin, and therefore the same extrinsic pose.

As each ultrasound recording may have a different FOV size and sampling (despite identical probe pose), it can be useful to first compute a common bounding box of all FOVs, into which the individual recordings are resampled. Furthermore, the target geometry can also be chosen independently from the US probe. For example, it can be derived from a calibrated CT (computed tomography) scan, in order to resample the data from one scan into the other. Lastly, the source geometry does not have to be enclosed completely by the target geometry.

Figure 8.4 – 3D acoustic grid for calibration of ultrasound transducer.
Parts: ① conical connector, ② swivel nut, ③ threaded rods, ④ 2D grids, ⑤ spacers,
⑥ stabilization ring. **(a)** Complete assembly. **(b)** Detail view of stacked 2D grids.
(c) Schematic drawing of individual 2D grid layer. Note the asymmetric grid structure.

8.3.4　3D acoustic calibration grid

The intrinsic transformation of the ultrasound probe is calibrated by the manufacturer. Its extrinsic pose with respect to the global coordinate system has to be calibrated, since the transducer can be positioned freely using the hydraulic arm. Additionally, the exact location of the local origin of the acoustic field of view within the probe remains unknown. A 3D acoustic calibration grid was designed and built (fig. 8.4), which can be placed inside the perfusion bath and attached to the motorized rotation stage. Hereby the extrinsic calibration can be performed in situ in the very same manner as with the optical calibration target. The transformation between the coordinate systems defined by the two calibration grids is given by a simple translation in z-direction, which can be deduced from the noted positions of the vertical translation frame during calibration acquisitions.

The 3D calibration grid is composed of four stacked 2D grid layers. At first, I thought about using nylon threads, since this has been used before [30, 78, 112]. Following a recommendation from our workshop, the layers have been manufactured with a 3D-printer, which made the assembly more easy and precise than weaving and spanning threads. The 1 mm thick layers are slid onto the threaded rods, separated by 19 mm long spacer tubes, and secured with nuts. All 2D grid layers are identical, but can be mounted rotated in steps of 45°. Further, note that the grid pattern of each layer is asymmetrical (see fig. 8.4c). These two aspects provide a symmetry-free 3D pattern, which has minimal obstruction and whose orientation to the transducer can be recognized uniquely. Despite its coarse spacing, the grid can be used not only for extrinsic calibration, but also for correction of the intrinsic calibration scaling due to different medium density.

8.3.5　Experimental protocol

For Langendorff perfusion experiments with simultaneous optical mapping and 4D ultrasound acquisition of a contracting heart, the experimental protocol needs to be extended with steps for setup and calibration of the acoustic equipment.

1. **Optical setup and calibration:** Insert optical calibration grid. Adjust pose and FOV of cameras. Note frame position. Rotate grid and take pictures. Take out grid.

2. **Acoustical setup:** Insert heart. Adjust pose and FOV of US probe. Note frame position.

3. **Experiment:** Load heart with fluorescent dye. Perform synchronous optical and acoustical recordings. Do not move US probe nor cameras. If frame is moved, note position.

4. **Heart shape reconstruction:** Add contraction blocking agent (e.g. Blebbistatin). Document heart shape with geometry camera. Take out heart.

5. **Acoustical calibration:** Insert 3D acoustic grid. Take US recording. Note frame position.

8.3.6 Calibration of transducer pose

The recording acquired in step 5 of the experimental protocol is used to calibrate the extrinsic pose of the ultrasound probe. An interactive Python application was developed for supervised calibration based on visual alignment of the measured echo data with a line representation (fig. 8.5a) of the calibration grid. First, the lines visual is positioned along the z-axis, such that it matches the calibration grid's global pose during acquisition, taking into account the vertical positions of the translation frame during acoustical and optical calibration. For manual calibration, the volume and the lines visuals are then drawn together. The volumes position is interactively translated, rotated and if necessary scaled, until a best match of the visible echos of the grid layers, rods, and rings with the respective features of the lines visual is achieved. The final calibration shown in (figs. 8.5b and 8.5c) was achieved in heated Tyrode's solution without scaling, i.e. using the unchanged intrinsic calibration of the manufacturer. For better visibility during the manual alignment process, the visualization can be limited to only show selected grid layers and surrounding volume data $\pm 10\,\mathrm{mm}$ (fig. 8.5c). The $4{\times}4$ matrix of the manually calibrated pose is saved and can be reused for all US scans of the heart, where the transducer was in the same position, regardless of the respective size of acoustical FOV.

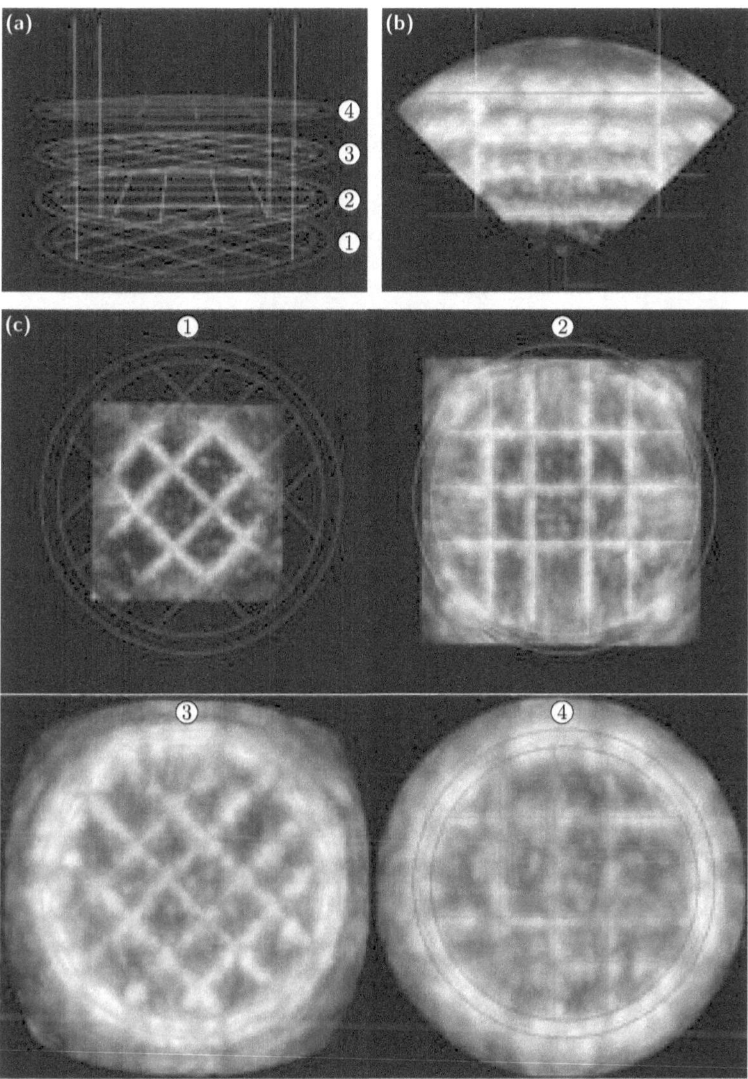

Figure 8.5 – Manual cxalibration of US transducer pose. (a) Perspective view of acoustic calibration grid drawn as colored lines visual, showing the grid layers ① to ④ and vertical rods. (b) Side view of simultaneous visualization of lines visual and resampled volume data of the acoustic calibration grid. Shown is the situation after manual calibration (see text). (c) Top views of individual grid layers and surrounding volume data.

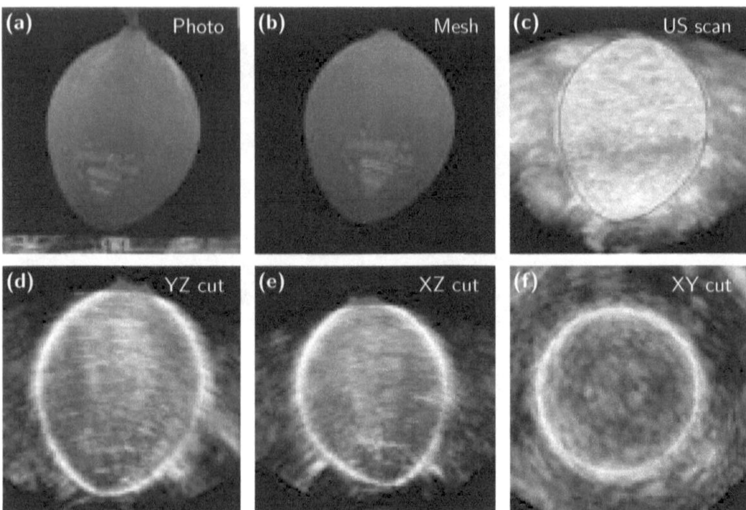

Figure 8.6 – Validation of ultrasound calibration. (a) Photo of water-filled balloon inside water-filled bath. (b) Textured mesh obtained with optical 3D shape reconstruction. (c) Calibrated US volume scan with outline of mesh (same view as panel b). (d)–(f) Cross sections of volume data (slice) depicted over cutaway drawings of mesh (half body), perpendicular to x, y, and z axis, respectively.

8.4   Validation

For validation of the methods, the simple geometry of a balloon was acquired with the US device and validated against an optical reconstruction thereof, which was done according to chapters 3 and 4 and has a very low reconstruction error. Balloon and bath were filled with cold tap water. For this case, the manufacturers intrinsic calibration had to be slightly corrected by a scaling factor of 0.946 during US probe pose calibration, in order to achieve a good match of the calibration grid lines visual and ultrasound echos. This was to be expected, as the density of tap water is lower than that of Tyrode's solution, which reduces the speed of sound, and therefore increases the estimated distances. Results are shown in fig. 8.6. The quality of the extrinsic calibration with intrinsic correction factor can be assessed from the cross sections (figs. 8.6d to 8.6f): The mesh intersects the fuzzy echos of the balloon's skin right in the middle.

8.5   Results

Figure 8.7 shows the result of simultaneous 3D panoramic optical mapping and 4D ultrasound acquisition of a pig heart, as published in [95]. The heart was locally stimulated on the epicardium of the left ventricle (LV) with a unipolar electrical pacing electrode. We employed a simplified version of our 3D electromechanical

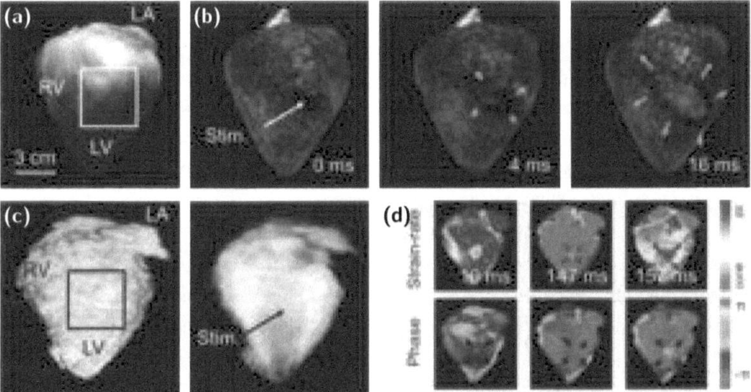

Figure 8.7 – Simultaneous optical and acoustical acquisition during local pacing.
Figure reused in part from [95]. **(a)** Optical 3D surface reconstruction of pig heart.
(b) Spread of action potential wave from local pacing site measured with 3D panoramic
optical mapping. **(c)** Solid and transparent volume visualization of heart acquired with
4D ultrasound. **(d)** Strain rate and phase computed from 4D volume speckle tracking.

optical mapping (3D-EMOM) technique. 2D optical motion tracking was applied to
videos of all four fluorescence cameras, as described in chapter 7 and [88]. Instead of
using this data for estimating the 3D motion of the epicardium, we directly warped
and projected the stabilized videos onto the static 3D mesh reconstructed after the
experiment. The focal origin and elliptical spread of the excitation wave can be
seen in both data sets, the optical mapped fast action potential (fig. 8.7h), and the
acoustical mapped following slow mechanical contraction (fig. 8.7j). Since these early
application results were obtained with visualization and calibration methods that
were still in development and did not yet achieve the quality shown in validation,
we did not evaluate the distance between the optical and acoustical early activation
sites quantitatively.

8.6 Discussion

Achievements. The viability of simultaneous, calibrated 4D ultrasound acquisition of a beating whole heart in a 3D panoramic optical mapping Langendorff-
perfusion setup was demonstrated. For this extent, a special perfusion bath with
integrated acoustic window was developed, which facilitates US measurements from
below without obstruction of the panoramic cameras. Additionally, a 3D acoustical
calibration grid was developed, which can be used to estimate the pose of the US
probe in situ without external tracking. A single short acquisition at high spatial
resolution is sufficient for documentation of the US probe's pose in the experiment.
Although the following calibration has to be performed manually, the procedure of
rotating and translating the volume until best match with the grid lines is achieved

is fast and straight forward. The very good match of calibrated optical surface reconstruction and calibrated US volume data was shown.

The combined optical and acoustical mapping setup allowed early application and results are part of a larger publication by Christoph et al. [95]. To our knowledge, this is the first time this combination has been used. We employed 4D ultrasound based strain-imaging, and provided experimental evidence of co-existing mechanical and electrical phase singularities of scroll waves.

Limitations and outlook. An automatic algorithmic optimization of the manually estimated transducer pose would provide more objective calibration results. However, given the amount of noise in the acoustic data, it remains to be seen if the calibration actually will be better than it already is.

Currently, the transformation between the coordinate systems established by optical and acoustical calibration relies on the common mounting point of the calibration grids at the motor. Small errors may result from imperfect manufacturing of the grids, and the unvalidated assumption that the motor's axis of rotation (which defines the z-axis of the optical CS) coincides with the movement axis of the vertical translation frame. While this is not a serious issue, it could be addressed by attaching optical markers to the 3D acoustic calibration grid, in order to estimate its pose with help of the cameras. Alternatively, a combined optical and acoustical calibration object could be devised for simultaneous calibration of cameras and ultrasound.

The spatial resolution is given by the ultrasound transducer and decreases with distance. The US volume is compromised by spurious echos, caused by reflections from the water surface and hard bath walls. In order to improve the signal to noise level and reduce spurious echos, ultrasound-absorbing materials could be introduced at the bottom of the bath and at water level without obstruction of the optical mapping. Also our choice of materials for the acoustic calibration grid may not be the best suited selection in terms of US reflection and absorption. The presented 3D grid is our first version which gave such good results, that we did not seek further improvement.

The presented in situ calibration method does not use live tracking of the transducer pose, and therefore requires it to be fixed during the experiment. If possibility of re-positioning or live tracking is desired, an external tracking system could be added with the established methods presented in section 8.2.

Here, our main purpose for use of the ultrasound system was the acquisition of the intramural motion of a contracting heart, therefore a 4D probe was used. The same transducer can of course be used for measurement of a stationary heart, where speed is not imperative and the scan can be performed at higher spatial resolution. Additionally, the stationary heart could be rotated with the motor and multiple scans be averaged, which should further reduce noise from spurious echos. The availability of an US machine at the experiment with support for multiple transducers would also allow other use cases. Using a 2D probe would allow high-resolution

3D scanning of a stationary heart, by an adapted freehand 3D US reconstruction method with a sweeping/rotating movement of the transducer or preferably rotation of the heart with the motor. Another use case is recording of a 2D slicing plane through the myocardium, as we have done in [95]. Currently, only one transducer can be placed under the bath. In order to switch between transducers without rearranging, a new bath could be built to incorporate a 2D probe in addition to the 4D probe.

Chapter 9

Multi-channel ECG

9.1 Introduction

This chapter covers the extension of the 3D panoramic optical mapping system with the option to simultaneously record several hundred electrocardiogram (ECG) signals for development and validation of advanced non-invasive diagnostic methods.

In the clinical application, conventional ECGs are used for non-invasive monitoring and assessment of the heart's current state. Electrodes are attached to the patients skin at specific sites on the torso and limbs and connected to a sensitive recording device. The electrical potential difference measured between any two electrodes is called *lead*, not to be confused with the physical electrode connection itself. The potential of a so-called *unipolar lead* is measured against a virtual electrode (Wilson's central terminal), which is the averaged potential measured by three limb electrodes attached to the right arm, the left arm, and the left foot. A conventional 12-lead ECG uses ten electrodes attached to the skin on a patient's torso and limbs in a standardized arrangement, which compose three bipolar limb leads and nine unipolar leads. When all cardiomyocytes are at rest, voltages measured across the body surface are zero. A depolarizing excitation wavefront locally increases the amount of negatively charged ions in the exctracellular medium, thereby creating a non-uniform distribution of charges. Each ECG lead measures a potential difference, which depends on the projection of the total electrical dipole moment onto the *vector* between the two respective electrodes. Amplitude and polarity of deflections visible in the ECG graph therefore depend on the direction of travel of an excitation wave in the cardiac muscle tissue. A trained physician can look at the printouts of the various leads and carefully compare and analyze the wave forms in order to detect signs for abnormal behavior. This requires experience and is sometimes referred to as the *art of ECG interpretation* [72]. Correct electrode placement is crucial and patient-specific electrophysiological factors (e.g. age) have to be considered. In order to reliably detect and diagnose specific syndromes, different electrode arrangements may have to be used. Increasing the number of electrodes theoretically provides more information, but in practice makes the interpretation more difficult for the human mind.

In order to assist the medical staff with this task and to give more insight into the heart's dynamic behavior, work on novel computational methods has been undergone for some decades, with a recent upsurge in academic and commercial interest [20, 37, 43, 98]. The aim of these efforts is a combined ECG and computer system, that uses a large number of electrodes attached densely to a patient's torso, and can non-invasively calculate and visualize the electrical activity of the heart in high spatial and temporal resolution, comparable to optical mapping or invasive techniques (fig. 9.1). Those methods go under different names in the literature, one of which is non-invasive electrocardiographic imaging (ECGI). A three-dimensional model of the patient's heart and torso, together with the electrode positions is obtained from a CT (computed tomography) or MRI (magnetic resonance imaging) scan. From that, a model of the electrically conductive volume between heart and body surface (lungs, ribs, muscles, ...) is derived as a mathematical description. Simplifying assumptions about the kind and distribution of current sources on the heart surface allow to numerically solve the inverse problem. Early ECGI systems allowed mapping of the epicardial surface only. Novel methods aim to reconstruct both, epi- and endocardial surfaces [83]. Clinical application of the ECGI methodology includes "pre-operative diagnosis of complex heart rhythm disturbances, therapy planning, monitoring as well as follow-up."[1] However, extensive validation of the output of the non-trivial computations is difficult to obtain with in vivo experiments, due to the necessity for open-chest surgery or usage of catheter electrodes. This complication limits the current understanding to what degree and under which circumstances the computed surface potentials reflect the real dynamics of the heart.

On the one hand, our 3D panoramic optical mapping setup provides an ideal base for validation of non-invasive electrocardiographic imaging methods with high spatiotemporal resolution. On the other hand, ECGI can be a good supplement to the other non-invasive multimodal techniques applied in this work, in order to visualize electrical activity on the endocardium, which is not easily accessible to optical mapping.

Methods and experiments presented in this chapter were developed and conducted as part of a Shared Expertise project between DZHK (Deutsches Zentrum für Herz-Kreislauf-Forschung e.V.) partner sites Göttingen and Hamburg/Kiel/Lübeck (Prof. Dr. med. Kuck, Asklepios St. Georg, Hamburg), together with external collaboration partner EP Solutions SA, Yverdon-les-Bains, Switzerland (EPS). Aim of the project was to compare and evaluate non-invasive electrocardiographic imaging against 3D panoramic optical mapping. However, results cannot be disclosed at this place.

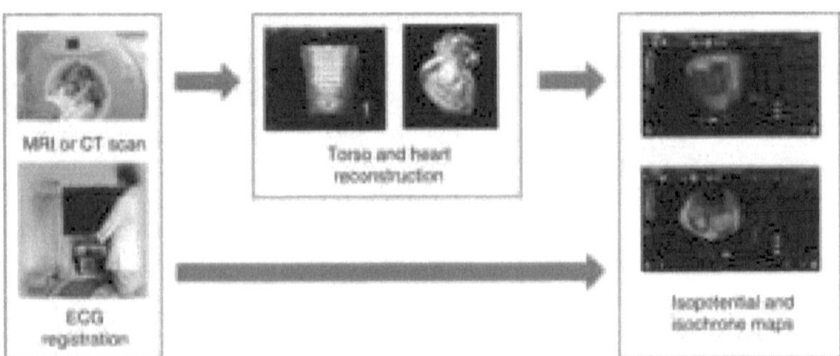

Figure 9.1 – Non-invasive endo- and epicardial electrophysiology study procedure. Figure reprodued unmodified from [83] under license CC BY-NC 4.0 [114].

9.2 Materials and methods

9.2.1 Electrode plates

Acrylic glass plates were designed to be inserted into the perfusion bath with acoustical window (cf. fig. 8.2d) and hold a total of 212 electrodes, covering all sides of the octagonal bath. The plates were built by our machine shop and fitted with electrodes by colleague Laura Diaz. Circular silver electrodes ($\varnothing = 5\,\mathrm{mm}$) were arranged on a grid pattern on four narrow and four wide plates, leaving a central opening on the wide plates for unobstructed optical mapping (fig. 9.2a). The electrodes were inserted in pre-drilled holes on the front side of the plates and soldered to wires from the back. The plates are fitted with pins at the bottom, which allow easy mounting in predefined positions, by inserting into matching holes in the bath's bottom and fixation of the plate against the glass windows at the top with custom-made clamps (fig. 9.2c). Plates can be removed and reinserted in the same position. The positions of the individual electrodes are therefore precisely known within the bath's *local* coordinate frame (fig. 9.2b).

9.2.2 Multi-channel ECG recording system

A commercial 224-channel ECG system[2] for noninvasive cardiac electrophysiology studies (amplifier, computer and recording software) was provided by the company EP Solutions. Electrical integration into our perfusion setup was done together with personnel of EPS and help of Laura Diaz and Andreas Barthels (both MPIDS). In order to reduce noise, the amplifier was powered by a 12 V car battery and the electrical bath heating turned off during recording. Top and bottom center electrodes on the plates facing the left and right ventricles were designated to take the role of the limb electrodes, i.e. their signals were used to form the Wilson's terminal, which served as the reference for all other unipolar leads. For synchronization with the

---

[2]AMYCARD 01 C, EP Solutions SA, Yverdon-les-Bains, Switzerland

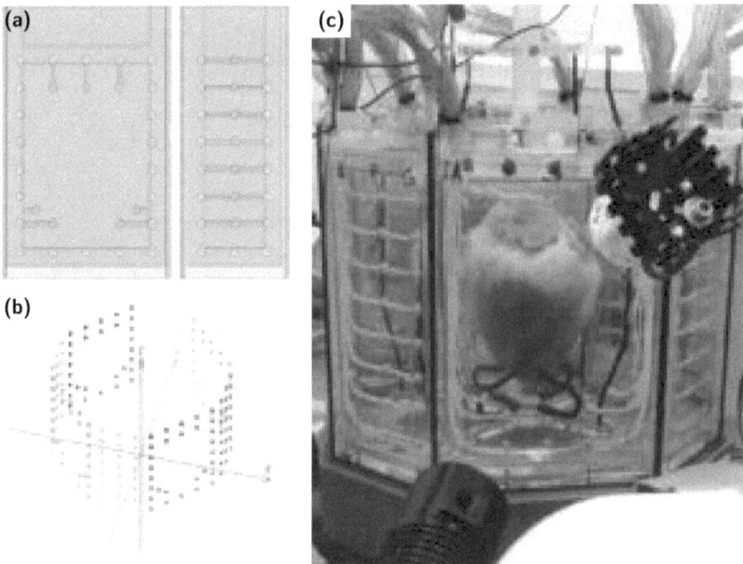

Figure 9.2 – Perfusion bath equipped with electrode plates for 212-channel ECG recording. (a) Design of wide and narrow electrode plates. (b) Positions of the electrodes within the local coordinate system of the bath. (c) Plates inserted into bath, facilitating optical mapping through wide windows.

optical mapping recordings, the trigger signal of the cameras was fed into one of the unused ECG channels, since the amplifier does not provide any other distinguished input or output. The TTL (transistor-transistor logic) signal was electrically isolated and its 5 V amplitude reduced, using a custom made optocoupler with 750:1 voltage divider.

9.2.3 Estimation of bath pose

For all other methods described in the previous chapters, the position of the perfusion bath does not need to be known. The only requirement for optical mapping is, that the bath position does not change during the experiment, since otherwise the cameras' calibrations are lost. For application of ECGI, the exact positions of bath, electrodes, and heart need to be known. The design of the electrode plates ensures that the position of the electrodes is known with high precision with respect to the local coordinate system of the bath. The heart is reconstructed with respect to the coordinate system defined by the motorized rotation stage. The motor was set up to be centered above the bath, which should align the z-axes of both coordinate systems, but it was later discovered that the mounting of the aluminum frame at that time was not perfectly stable (no diagonal beam yet), causing it to move slightly from experiment to experiment unnoticed.

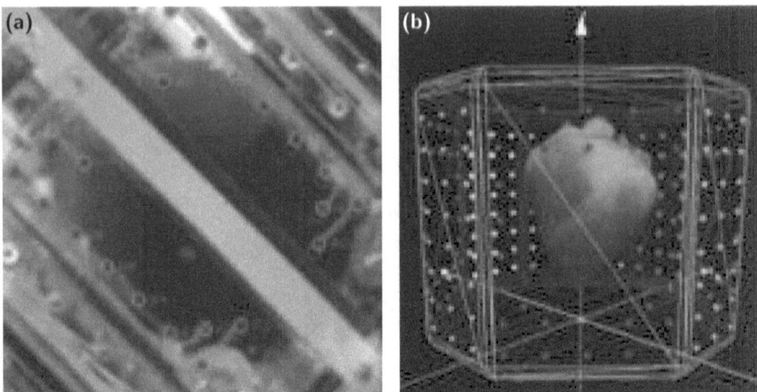

Figure 9.3 – Estimation of bath pose from electrodes visible in calibration images.
(a) Camera 3 looking onto the inside of electrode plate I, showing manually determined
2D image coordinates of identifiable electrodes (red) and positions of the 3D electrode
positions projected to this camera view before (blue) and after (green) pose estimation.
(b) Visualization of reconstructed heart mesh, electrode positions, and conductive liquid
volume (grey wireframe) in coordinate system of bath.

To correct for this error, we found that it is possible to obtain the transformation
between both coordinate systems directly from the calibration images of the four
fluorescence cameras, as some of the electrodes can be best seen in the foreground
and background of those images, when the calibration target plate is perpendicular
to the image plane (fig. 9.3a). Individual electrodes were identified and their center
points in image coordinates were determined by manually fitting circles. Only the
electrodes in the background were used, as their position could be determined more
reliably, because the cameras were looking on the front sides of the electrode plates.
The pose of the rigid assembly of bath and electrode plates was estimated using
a Levenberg-Marquardt routine, by variation of a three-dimensional rotation and
translation until the reprojection error (mean squared distances of projected and
measured electrode positions for all four cameras) became minimal. The resulting
transformation was used to bring all components into the same frame of reference
(fig. 9.3b).

9.2.4 ECGI validation study

Experiments for validation of the ECGI method against optical mapping were done
using nine hearts of Göttinger Minipigs. The general protocol followed a typical 3D
panoramic optical mapping experiment, where the heart is immobilized using Blebbistatin and additionally supported with a stabilization frame. The main objective
of the experiments was to record different dynamic states of the heart: 1) sinus
rhythm, 2) tachycardia, 3) ventricular fibrillation, 4) pacing from different locations
on the ventricles. Not all states could be realized with all hearts.

Experiments were conducted with personnel from Research Group Biomedical Physics and EP Solutions. The Multi-ECG recording system was monitored and operated by personnel of EPS. In order to guarantee a blinded study design, only BMP personnel was in the room during step 4, who positioned the pacing electrode. Additionally, the recorded video data was not disclosed to EPS.

9.2.5 Modeling of heart and conductive liquid volume

The 3D geometrical shape of the heart surface was reconstructed with our usual Shape from Contour toolchain. For each heart, two triangular meshes of different spatial resolution were generated by remeshing with Isoparameterization at different sampling rates, as explained in section 4.2.5. A high-resolution mesh was used for optical mapping, whereas a mesh with lower resolution (about 2000 to 2500 vertices) was provided to EP Solutions for solving the inverse ECGI problem. The volume of liquid perfusion solution between electrode plates was modeled from the CAD drawings (fig. 9.3b). Sharp edges were rounded to a radius of 10 mm for computational reasons. To obtain the endocardial geometry, each heart was preserved after the experiment and a CT scan performed as described in chapter 6.

9.2.6 Analysis

The CT volume data was supplied to EPS together with the meshes of heart surface and surrounding liquid as well as electrode positions. ECG data for each optical recording was cut to beginning and end of camera trigger series in the corresponding ECG channel. ECGI reconstruction was performed on the data at original sampling rate (1000 Hz) by EPS with method of their choice. Resulting surface potential timeseries for every vertex of the mesh were resampled to match the frame rate of the cameras (500 Hz) and mapped to polar texture videos with spatial interpolation. 3D panoramic optical mapping videos of sinus rhythm, ventricular tachycardia, and ventricular fibrillation were processed and analyzed with methods of chapter 5. In order to be comparable, ECGI reconstructions were analyzed with the same tools.

Experiments of pacing at different locations on the ventricles were performed to assess the spatial resolution and accuracy of ECGI reconstruction. Optical early activation sites were determined in the video textures by hand and then converted to 3D coordinates on the heart surface. Electrical early activation sites were determined by a physician (excluding stimulation artifact) and also stored as 3D points. Differences between the corresponding optically and electrically determined locations were analyzed.

Due to delays, results unfortunately cannot be shown yet.

9.3 Discussion

In retrospect, the ECGI validation project was started too early, before the 3D optical mapping setup and software was thoroughly tested, which led to unforeseen problems caused by still present mechanical instabilities and undetected software bugs that surfaced late in the analysis. It took a lot of time figuring out what went wrong and how to estimate and compensate the errors in order to save and analyze the acquired experimental data.

We were lucky that the available calibration images allowed the ad-hoc solution for estimation of the bath pose. Future experiments should incorporate other kinds of markers on the electrode plates or the bath itself, that allow automatic detection, e.g. (Ch)ArUco markers [79].

The usefulness of experiments with combined optical mapping and multi-channel ECG recording is underlined by a demonstration of Bear et al. [106], who developed an ex vivo perfusion system with a whole heart inside a torso shaped tank featuring 128 embedded electrodes and window for monocular imaging. The setup was used for validation of surface potentials reconstructed by ECGI with optically mapped membrane potentials together with direct measurements obtained from sock electrodes. While I do not know enough about the ECGI method itself to judge their findings and the specific reasons to use a torso-like shape for the tank, I speculate that this shape is not that important, as long as the inhomogeneous electrical conductivity of the internal tissues (e.g. lung, ribs, muscles, skin) is not modeled at all. I would therefore argue, that the benefits of our multi-camera setup with 3D panoramic optical mapping and the availability of simultaneous recording of the other modalities will be of more value for the advancement of ECGI than a realistic geometry of the chest surface.

Meanwhile, my colleagues have developed and build a new 256-channel ECG amplifier and recording device for our group, on the basis of a National Instruments PXIe-1085 system.[3] It yields recordings with good signal-to-noise ratio (SNR), even with small rabbit hearts inside the large bath with acoustical window.

Apart from validation, advancement, and eventual application of the ECGI methodology in our ex vivo experiments, the multi-channel ECG system is also useful for other research topics such as evaluation of complexity measures of far-field ECG signals (e.g. permutation entropy) and their angular dependencies [103, 104] in comparison to excitation patterns visible intramuraly and on the surface.

---

[3]3-HE-PXI-Express-Chassis with NI PXIe-8880, Xeon 8-Core-Controller Labview RT and FPGA Module PXIe-7846R R Series Multifunction I/O.

Part III

Conclusion

Chapter 10

Conclusion

The heart is a dynamical system, whose function as a pump depends strongly on the correct interplay of structural and dynamical factors. Many diseases such as cardiac death or cardiac infarction can be explained as a change of the spatio-temporal dynamics of the electro-mechanical system. It is not sufficient to study the respective processes isolated from another or at levels of molecular, cellular, and tissue scale. To better comprehend the spatio-temporal dynamics, it is imperative to capture it in its entirety on the whole organ — be it healthy or diseased. This shall also serve the development of novel therapeutic methods for control of cardiac dynamics with precise, temporally and spatially localized stimuli.

The goal of this work was the development and advancement of experimental and numerical tools in an ex vivo setup, in order to measure the key characteristics of the heart with high spatial and temporal resolution and make them accessible to basic research. For this purpose, an ex vivo Langendorff perfusion setup for isolated rabbit and pig hearts was enhanced using a combination of non-invasive techniques, to facilitate simultaneous, dense, and comprehensive multimodal measurement of structure, movement and electrical dynamics of a beating heart, on its surface and within. Optical techniques that reconstruct, map, and track the deforming heart surface were accompanied with acoustical, electrical, and radiographical measurement methods which image the bulk tissue. Table 10.1 gives an overview of how the individual methods complement one another.

In the following, I will summarize the main developments and results, and discuss the value for the scientific community. The last section gives an outlook for further improvements and applications.

		3D shape reconstruction	3D panoramic opt. mapping	3D surface deformation tracking	4D ultrasound	Multi-channel ECG	CT scanning
Structure	Domain:	Surface			Bulk		Bulk
	Resolution:	High			Low		Very high
	Where:	In situ			In situ		Ex situ
	Contraction:	Stationary			Beating		Stationary
Motion	Domain:			Surface	Bulk		
	Resolution:			High	Low		
Electrical dynamics	Domain:		Surface		Bulk	Bulk[**]	
	Resolution:		High		Low	Low	
	How:		Direct		Indirect[*]	Indirect	

Table 10.1 – Overview of the non-invasive methods combined in this work, listing their individual capacity to measure structure, motion, and electrical dynamics of a (beating) heart. Altogether, they form a comprehensive multimodal imaging system.

[*] Via inverse reconstruction of local membrane potential from following contraction.
[**] ECG measures electrical potential difference along a vector between leads through the heart. ECGI reconstructs the surface potentials on the epi- and endocardial surfaces.

10.1 Summary of major developments and achievements

Chapter 2: Ex vivo optical mapping of isolated hearts. The existing experimental setup for 3D panoramic optical mapping was improved and the mechanical components largely re-build, in order to facilitate rigid and accurate (re-)positioning of the bath, cameras, and heart, while being easily adjustable and flexibly usable to allow other experiments with less strict requirements. A motorized rotation stage with fluid feed-through and accessories adapter was incorporated into the perfusion system for attachment of and automated photography of the heart from multiple angles, as well as optical and acoustical calibration objects.

Chapter 3: Camera calibration. Building upon a standard camera model and calibration technique with chessboard patterns, a panoramic multi-camera calibration method was developed, which uses simple-to-manufacture flat calibration targets that are rotated by the motorized stage. High spatial accuracy is achieved by incorporation of multiple views and automatic misalignment compensation in the extrinsic calibration routine. Reprojection errors of about 0.03 mm to 0.04 mm are obtained with all cameras for imaging of both, rabbit and pig hearts. Additional analysis of diffraction effects and camera model distortion parameters revealed the feasibility of inclined camera positioning with respect to the bath windows in future applications.

Chapter 4: Static 3D heart shape reconstruction. The toolchain used for three-dimensional geometry reconstruction of the hearts epicardial surface is based on established techniques, which were advanced towards automation, retaining points for manual intervention. The main Shape from Contour (SfC) algorithm uses silhouette images of the static heart acquired under various viewing angles. Different segmentation techniques for generation of the silhouette images were tested. A spherical surface parameterization technique was used for generation of unique texture coordinates of the whole heart, especially at the wrinkly atrial regions. The high accuracy of the reconstruction was validated with a static test object (reconstruction error $\leq 0.35\,\text{mm}$). As an alternative to the toolchain, reconstruction method Structure from Motion (SfM) was briefly evaluated.

Chapter 5: 3D panoramic optical mapping. Projection and blending of camera images was implemented in OpenGL for fast and lossless processing of 16bit original videos. Most standard analysis methods for excitation wave dynamics can be directly applied to the 2D polar surface texture maps. Three-dimensional analysis on the curved surface can be formulated as a quasi-2D methods through storing of geometrical properties of the irregular triangular mesh in 2D maps, e.g. pre-computation of push-forward tensors for calculation of apparent conduction velocity.

Chapter 6: CT scanning. A special sample container was built, which enables a CT (computed tomography) scan of the heart in a hanging position. This way, the deformation of the heart shape after the experiment is minimized, which allows a realistic high-resolution documentation of the heart's internal anatomy. For instance, whole pig hearts can be scanned at $0.0564\,\text{mm}$ resolution.

Chapter 7: 3D surface deformation tracking. A novel marker-free, purely image-based 3D surface deformation tracking was developed, by combination of two existing methods, namely above high-resolution static surface reconstruction and 2D motion tracking using contrast enhanced original fluorescent videos. The static 3D shape template was deformed according to the 2D displacement vector fields computed for each camera individually. Hereby, feature matching between neighboring cameras becomes obsolete, which is necessary for traditional stereoscopic triangulation, but fails for large baseline separation. Additionally, an unexpected new method for light field estimation was devised, which allows significant reduction of lighting-induced motion artifacts without ratiometric imaging, using the moving heart itself as a probe. The methods were demonstrated in proof of concept experiments with rabbit hearts, using four cameras positioned with very large baseline separation at 45° intervals, covering half of the heart [88]. More cameras can be incorporated without further ado, permitting full 360° deformation tracking.

Chapter 8: 4D ultrasound. A 4D ultrasound probe was integrated into the setup looking up to the heart through an acoustic window in the base of the bath, granting obstruction-free panoramic optical mapping with simultaneous ultrasonic volume acquisition. While the geometry of the scanning field of view (FOV) for a particular recording is known from the acquisition settings and manufacturer's specifications, the pose of the transducer was determined with the help of a 3D calibration grid that can be attached to the motorized rotation stage. This simple in situ calibration technique yields direct congruence between the optical and acoustical coordinate systems and measured data. Early application of the simultaneous optical and acoustical measurement methods demonstrate the potential for future research, as we found evidence for co-localized electrical and mechanical phase singularities/filaments [95].

Chapter 9: Multi-channel ECG. Multi-electrode plates for all eight sides of the bath were built that hold a total of 212 electrodes, with four openings for obstruction free panoramic optical mapping. A multi-channel amplifier and recording system was integrated for synchronous electrocardiogram (ECG) acquisition. Estimation of the bath and electrodes pose with respect to the cameras' coordinate system completed the calibration of the mechanical components. This provided the grounds for advanced experiments for validation of non-invasive electrocardiographic imaging (ECGI).

Software. A large part of the work of this PhD project was the advancement and development of the necessary software tools for:

- Control of rotation stage and cameras for camera calibration, heart shape documentation, and fluorescence recording.
- Multi-camera calibration routine with misalignment compensation. Object pose estimation from multiple views.
- Image segmentation, 3D reconstruction, surface parameterization, image projection, and 3D visualization of heart surface mesh.
- 3D heart surface deformation tracking from multiple 2D displacement fields with static reference mesh.
- Ultrasound (US) volume visualization, interactive calibration, and resampling between different calibrated geometries.

These tools are based on a number of free open source software libraries and will itself be made available as such, to lower the entry barrier for other researchers.

10.2   Discussion and outlook

Value of the setup and methods as a whole. Table 10.1 lists the different methods and gives an overview of their individual capacities to capture the structure, motion and electrical dynamics on the heart's surface and within. Each property of interest is measured in two or more ways, albeit at different resolutions and different locations. The strength of the system lies in the combination and mutual supplementing of these individually limited methods. In this sense, the whole is greater than the sum of its parts. While there is —as always— still work to be done, the foundations are laid: The methods work well together, forming one of, if not *the* most versatile and comprehensive setup for ex vivo multimodal imaging in the community.

In this book I tried to be thorough in listing all important advantages of the developed methods, as well as their challenges and limitations, in order to provide a good starting point for researchers wanting to employ one or more of the techniques in their own experiments. Among other things, delays occurred due to moving of the experimental setup between laboratories of different buildings, and due to software errors, that went unnoticed for a long time and were unnervingly hard to track. Ultimately, these delays prevented the conduction of experiments in this work, where all techniques are truly employed simultaneously on a single heart. Nonetheless, I do not see any obstacles for doing so in the very near future.

Consolidation of protocols for experimental application. The methods for multimodal measurement developed in this work are ready to be applied all at once in combined experiments. In the near future, the setup shall be used within our group as it is on a regular basis to address current research questions. Therefore, the consolidation and documentation of best-practice experimental protocols and analysis steps will have top priority for training of colleagues and subsequent publication. It is planned to release the codes as a full-featured open source software suite for multimodal measurement and analysis, together with construction drawings and exemplary original data and analysis results. In the longer run, several ways for further improvement of the numerical and experimental methods are possible.

Advancement of numerical methods. It should be possible to improve the quality of the lightfield estimation for 3D surface deformation tracking, using longer and more diverse motion sequences of non-periodic activity for a denser spatial sampling of the lightfield. Validation of 3D surface deformation tracking can be done on the basis of ground truth test data generated with computer graphics. In this way, also different numbers of cameras and various arrangements around the bath can be evaluated.

The usage of SfM as an alternative method for static 3D shape reconstruction was already evaluated in section 4.4. Maybe Non-rigid Structure from Motion (NrSfM) [99] can be used as another alternative shape reconstruction technique, that does not require the heart to be immobilized. An even greater advancement would be, if NrSfM could also be applied for 3D surface deformation tracking.

In the current state, the experimental setup permits simultaneous acquisition of structure, motion and electrical activity. In this work the methods have been used mostly independently alongside each other. Further work needs to be done to unlock the full potential of the setup by conjoint analysis of the generated multimodal datasets for reconstruction, tracking, and signal analysis. Using data assimilation techniques such as the Kalman filter [53], fusion and cross improvement of optical

3D surface deformation tracking with motion tracking of 4D US recordings should be possible, with the help of the high-resolution structural model obtained by CT scanning. The ultimate goal would incorporate all measured data in a detailed electro-mechanical model with realistic 3D geometry.

Extension of experimental setup. Further improvements of the setup will involve use of another geometry camera and better white light illumination. Use of additional cameras will allow 360° whole heart optical mapping and surface deformation tracking without further ado. Thanks to the 3D lightfield estimation, the additional cameras can be used for simultaneous multi-wavelength fluorometry, e.g. mapping of intracellular Ca^{2+} concentration dynamics. Nonetheless, established ratiometric methods can be easily incorporated, if desired. This is not limited by technical obstacles, but only by cost for the equipment.

Research questions in the background. Above considerations point out the current value of the multimodal mapping system for cardiological research, as well as its potential for further developments. There is a large number of scientific problems that will benefit from comprehensive multimodal and three-dimensional imaging methods. Just to name a few: the study of electromechanical dynamics of the healthy heart as well as in various disease models; complexity of excitation patterns visible intramuraly vs. on the surface vs. in far field ECG; development of novel strategies for cardiac control and defibrillation; validation of techniques for inverse reconstruction of electrical activity from ECG and US measurements; mechanisms of excitation-contraction coupling and feedback. I am convinced that experimental setups for multimodal imaging, such as this one, will enable a better understanding of the mechanisms underlying cardiac diseases and pave the way for novel therapies.

Part IV

Appendices

Appendix A

Validation of camera calibration

The tests were done as follows. First, 72 images of a calibration target rotating about the vertical z-axis were rendered, using standard OpenGL (Open Graphics Library) transformation pipeline with usual perspective camera. For this, the image size (1584 px × 2376 px), and view and projection matrices of the virtual camera were set up to mimic the actual DSLR camera, and the virtual 3D model of the calibration target was textured with the same chessboard patterns on front and back that would be used for pig hearts. Second, the images were artificially warped according to the respective distortion coefficients. Third, chessboard corners were searched in the (possibly distorted) images and calibration procedure was performed. Fourth, images were undistorted according to calibrated intrinsic distortion coefficients. Fifth, rendering of the scene again, now using OpenGL view and projection matrices which were derived from the calibrated model as explained in section 3.2.4 above. Images generated in steps four and five were compared to the original renderings and the calibrated camera parameters and pattern pose matrix compared to the initial values.

Detailed listings of all the values of the ground-truth matrices as well as the calibrated camera model parameters and reconstructed OpenGL matrices are given in the following tables A.1 to A.18. The left column of these tables shows all the values used for rendering the ground truth images. The middle column shows in the top row 'parameters' the constraints of the camera model parameters during calibration, and whether the misalignment estimation was performed during extrinsic calibration. Resulting values of the calibration are listed in the rows below and can be compared to the ground truth values in the left column. Instead of comparing the values of the calibrated camera matrix K and pose $(R|t)$, it is better to compare the values of the OpenGL projection and view matrices P and V shown in the right column, which were derived from the calibrated model as explained in section 3.2.4. Note, that the near and far clipping values in projection matrix $P_{\mathrm{cal.}}$ do not match those of projection matrix P used for rendering of the ground truth images. This is because z_{near} and z_{far} cannot be derived from the camera matrix K. Instead they have to be either set manually, or as in this case chosen automatically, such that the model is within the clipping range. The image(s) show the rendered calibration target before (and after) simulated distortion, using the ground-truth values. Renderings obtained from calibrated values are virtually indistinguishable and therefore not shown.

	Ground truth	Calibrated camera model	Validation
Parameters	width $= 1584$, height $= 2376$ $\text{fov}_y = 18.5°$ $c_x = 791.5,\ c_y = 1187.5$	$f_x \overset{!}{=} f_y,\ (c_x, c_y) \overset{!}{=}$ image center, $p_1 \overset{!}{=} 0,\ p_2 \overset{!}{=} 0,\ k_1 \overset{!}{=} 0,\ k_2 \overset{!}{=} 0,\ k_3 \overset{!}{=} 0,$ Estimate misalignment: False	
Projection	$P = \begin{pmatrix} 9.2103 & 0 & 0 & 0 \\ 0 & 6.1402 & 0 & 0 \\ 0 & 0 & -2.5 & -1050 \\ 0 & 0 & -1 & 0 \end{pmatrix}$	$K = \begin{pmatrix} 7294.3 & 0 & 792 \\ 0 & 7294.3 & 1188 \\ 0 & 0 & 1 \end{pmatrix}$	$P_{\text{cal.}} = \begin{pmatrix} 9.21 & 0 & -6.31\mathrm{e}^{-4} & 0 \\ 0 & 6.14 & 4.21\mathrm{e}^{-4} & 0 \\ 0 & 0 & -1.2857 & -297.13 \\ 0 & 0 & -1 & 0 \end{pmatrix}$
Camera pose	$V = \begin{pmatrix} 1 & 0 & 0 & 0 \\ 0 & 0 & 1 & 0 \\ 0 & -1 & 0 & -520 \\ 0 & 0 & 0 & 1 \end{pmatrix}$	$(R\|t) = \begin{pmatrix} 1 & -6.79\mathrm{e}^{-5} & 1.48\mathrm{e}^{-6} & -0.035596 \\ 1.47\mathrm{e}^{-6} & -6.80\mathrm{e}^{-5} & -1 & -0.035764 \\ 6.79\mathrm{e}^{-5} & 1 & -6.80\mathrm{e}^{-5} & 519.98 \end{pmatrix}$	$V_{\text{cal.}} = \begin{pmatrix} 1 & -6.79\mathrm{e}^{-5} & 1.48\mathrm{e}^{-6} & -0.035596 \\ -1.47\mathrm{e}^{-6} & 6.80\mathrm{e}^{-5} & 1 & 0.035764 \\ -6.79\mathrm{e}^{-5} & -1 & 6.80\mathrm{e}^{-5} & -519.98 \\ 0 & 0 & 0 & 1 \end{pmatrix}$
Distortion	$k_1 = 0 \quad k_4 = 0$ $k_2 = 0 \quad k_5 = 0$ $k_3 = 0 \quad k_6 = 0$ $p_1 = 0 \quad s_1 = 0$ $p_2 = 0 \quad s_2 = 0$ $\tau_x = 0 \quad s_3 = 0$ $\tau_y = 0 \quad s_4 = 0$	$k_1 = 0 \quad k_4 = 0$ $k_2 = 0 \quad k_5 = 0$ $k_3 = 0 \quad k_6 = 0$ $p_1 = 0 \quad s_1 = 0$ $p_2 = 0 \quad s_2 = 0$ $\tau_x = 0 \quad s_3 = 0$ $\tau_y = 0 \quad s_4 = 0$	
Pattern pose	$M = \begin{pmatrix} 1 & 0 & 0 & 0 \\ 0 & 1 & 0 & -7.5 \\ 0 & 0 & 1 & 0 \\ 0 & 0 & 0 & 1 \end{pmatrix}$	$M_{\text{cal.}} = \begin{pmatrix} 1 & 0 & 0 & 0 \\ 0 & 1 & 0 & -7.5 \\ 0 & 0 & 1 & 0 \\ 0 & 0 & 0 & 1 \end{pmatrix}$ Intrinsic calibration: Repr. error $= 0.04326$ px, views $= 31$ Extrinsic calibration: Repr. error $= 0.041579$ px, views $= 31$	

Table A.1 – Test #0: Rendering of ground truth images with target centered in view, no distortion, no pattern misalignment. Intrinsic calibration with focal lengths forced equal, principal point forced centered, distortion coefficients forced zero. No estimation of pattern misalignment.

	Ground truth	Calibrated camera model	Validation
Parameters	width = 1584, height = 2376 $\text{fov}_y = 18.5°$ $c_x = 791.5,\ c_y = 1187.5$	$f_x \overset{!}{=} f_y,$ $p_1 \overset{!}{=} 0,\ p_2 \overset{!}{=} 0,\ k_1 \overset{!}{=} 0,\ k_2 \overset{!}{=} 0,\ k_3 \overset{!}{=} 0,$ Estimate misalignment: False	
Projection	$P = \begin{pmatrix} 9.2103 & 0 & 0 & 0 \\ 0 & 6.1402 & 0 & 0 \\ 0 & 0 & -2.5 & -1050 \\ 0 & 0 & -1 & 0 \end{pmatrix}$	$K = \begin{pmatrix} 7294.3 & 0 & 791.37 \\ 0 & 7294.3 & 1187.5 \\ 0 & 0 & 1 \end{pmatrix}$	$P_{\text{cal.}} = \begin{pmatrix} 9.21 & 0 & 1.68\mathrm{e}^{-4} & 0 \\ 0 & 6.14 & 8.70\mathrm{e}^{-6} & 0 \\ 0 & 0 & -1.2857 & -297.13 \\ 0 & 0 & -1 & 0 \end{pmatrix}$
Camera pose	$V = \begin{pmatrix} 1 & 0 & 0 & 0 \\ 0 & 0 & 1 & 0 \\ 0 & -1 & 0 & -520 \\ 0 & 0 & 0 & 1 \end{pmatrix}$	$(R\mid t) = \begin{pmatrix} 1 & 1.73\mathrm{e}^{-5} & -5.16\mathrm{e}^{-8} & 9.44\mathrm{e}^{-3} \\ -5.15\mathrm{e}^{-8} & -1.47\mathrm{e}^{-6} & -1 & -8.47\mathrm{e}^{-4} \\ -1.73\mathrm{e}^{-5} & 1 & -1.47\mathrm{e}^{-6} & 519.98 \end{pmatrix}$	$V_{\text{cal.}} = \begin{pmatrix} 1 & 1.73\mathrm{e}^{-5} & -5.16\mathrm{e}^{-8} & 9.44\mathrm{e}^{-3} \\ 5.15\mathrm{e}^{-8} & 1.47\mathrm{e}^{-6} & 1 & 8.47\mathrm{e}^{-4} \\ 1.73\mathrm{e}^{-5} & -1 & 1.47\mathrm{e}^{-6} & -519.98 \\ 0 & 0 & 0 & 1 \end{pmatrix}$
Distortion	$k_1 = 0 \quad k_4 = 0$ $k_2 = 0 \quad k_5 = 0$ $k_3 = 0 \quad k_6 = 0$ $p_1 = 0 \quad s_1 = 0$ $p_2 = 0 \quad s_2 = 0$ $\tau_x = 0 \quad s_3 = 0$ $\tau_y = 0 \quad s_4 = 0$	$k_1 = 0 \quad k_4 = 0$ $k_2 = 0 \quad k_5 = 0$ $k_3 = 0 \quad k_6 = 0$ $p_1 = 0 \quad s_1 = 0$ $p_2 = 0 \quad s_2 = 0$ $\tau_x = 0 \quad s_3 = 0$ $\tau_y = 0 \quad s_4 = 0$	
Pattern pose	$M = \begin{pmatrix} 1 & 0 & 0 & 0 \\ 0 & 1 & 0 & -7.5 \\ 0 & 0 & 1 & 0 \\ 0 & 0 & 0 & 1 \end{pmatrix}$	$M_{\text{cal.}} = \begin{pmatrix} 1 & 0 & 0 & 0 \\ 0 & 1 & 0 & -7.5 \\ 0 & 0 & 1 & 0 \\ 0 & 0 & 0 & 1 \end{pmatrix}$ Intrinsic calibration: Repr. error = 0.043193 px, views = 31 Extrinsic calibration: Repr. error = 0.041585 px, views = 31	

Table A.2 – Test #1: Same as test #0, but with free calibration of principal point.

Ground truth	Calibrated camera model	Validation

Parameters

width $= 1584$, height $= 2376$

$\text{fov}_y = 18.5°$

$c_x = 791.5,\ c_y = 1187.5$

$p_1 \overset{!}{=} 0,\ p_2 \overset{!}{=} 0,\ k_1 \overset{!}{=} 0,\ k_2 \overset{!}{=} 0,\ k_3 \overset{!}{=} 0,$

Estimate misalignment: False

Projection

$$P = \begin{pmatrix} 9.2103 & 0 & 0 & 0 \\ 0 & 6.1402 & 0 & 0 \\ 0 & 0 & -2.5 & -1050 \\ 0 & 0 & -1 & 0 \end{pmatrix}$$

$$K = \begin{pmatrix} 7294.4 & 0 & 791.37 \\ 0 & 7294.3 & 1187.5 \\ 0 & 0 & 1 \end{pmatrix}$$

$$P_\text{cal.} = \begin{pmatrix} 9.2101 & 0 & 1.68\mathrm{e}^{-4} & 0 \\ 0 & 6.14 & 9.06\mathrm{e}^{-6} & 0 \\ 0 & 0 & -1.2857 & -297.13 \\ 0 & 0 & -1 & 0 \end{pmatrix}$$

Camera pose

$$V = \begin{pmatrix} 1 & 0 & 0 & 0 \\ 0 & 0 & 1 & 0 \\ 0 & -1 & 0 & -520 \\ 0 & 0 & 0 & 1 \end{pmatrix}$$

$$(R|t) = \begin{pmatrix} 1 & 1.73\mathrm{e}^{-5} & -5.16\mathrm{e}^{-8} & 9.44\mathrm{e}^{-3} \\ -5.16\mathrm{e}^{-8} & -1.71\mathrm{e}^{-6} & -1 & -8.38\mathrm{e}^{-4} \\ -1.73\mathrm{e}^{-5} & 1 & -1.71\mathrm{e}^{-6} & 519.98 \end{pmatrix}$$

$$V_\text{cal.} = \begin{pmatrix} 1 & 1.73\mathrm{e}^{-5} & -5.16\mathrm{e}^{-8} & 9.44\mathrm{e}^{-3} \\ 5.16\mathrm{e}^{-8} & 1.71\mathrm{e}^{-6} & 1 & 8.38\mathrm{e}^{-4} \\ 1.73\mathrm{e}^{-5} & -1 & 1.71\mathrm{e}^{-6} & -519.98 \\ 0 & 0 & 0 & 1 \end{pmatrix}$$

Distortion

$k_1 = 0 \quad k_4 = 0$
$k_2 = 0 \quad k_5 = 0$
$k_3 = 0 \quad k_6 = 0$
$p_1 = 0 \quad s_1 = 0$
$p_2 = 0 \quad s_2 = 0$
$\tau_x = 0 \quad s_3 = 0$
$\tau_y = 0 \quad s_4 = 0$

$k_1 = 0 \quad k_4 = 0$
$k_2 = 0 \quad k_5 = 0$
$k_3 = 0 \quad k_6 = 0$
$p_1 = 0 \quad s_1 = 0$
$p_2 = 0 \quad s_2 = 0$
$\tau_x = 0 \quad s_3 = 0$
$\tau_y = 0 \quad s_4 = 0$

Pattern pose

$$M = \begin{pmatrix} 1 & 0 & 0 & 0 \\ 0 & 1 & 0 & -7.5 \\ 0 & 0 & 1 & 0 \\ 0 & 0 & 0 & 1 \end{pmatrix}$$

$$M_\text{cal.} = \begin{pmatrix} 1 & 0 & 0 & 0 \\ 0 & 1 & 0 & -7.5 \\ 0 & 0 & 1 & 0 \\ 0 & 0 & 0 & 1 \end{pmatrix}$$

Intrinsic calibration:
Repr. error $= 0.043191$ px, views $= 31$

Extrinsic calibration:
Repr. error $= 0.041587$ px, views $= 31$

Table A.3 – Test #2: Same as test #0, but with free calibration of focal lengths and principal point.

125

	Ground truth	Calibrated camera model	Validation
Parameters	width $= 1584$, height $= 2376$ fov$_y = 18.5°$ $c_x = 749.5$, $c_y = 1099.5$	$f_x \overset{!}{=} f_y$, $p_1 \overset{!}{=} 0$, $p_2 \overset{!}{=} 0$, $k_1 \overset{!}{=} 0$, $k_2 \overset{!}{=} 0$, $k_3 \overset{!}{=} 0$, Estimate misalignment: False	
Projection	$P = \begin{pmatrix} 9.2103 & 0 & 0.05303 & 0 \\ 0 & 6.1402 & -0.074074 & 0 \\ 0 & 0 & -2.5 & -1050 \\ 0 & 0 & -1 & 0 \end{pmatrix}$	$K = \begin{pmatrix} 7294.3 & 0 & 749.49 \\ 0 & 7294.3 & 1099.5 \\ 0 & 0 & 1 \end{pmatrix}$	$P_{\text{cal.}} = \begin{pmatrix} 9.2099 & 0 & 0.053038 & 0 \\ 0 & 6.1399 & -0.074057 & 0 \\ 0 & 0 & -1.2857 & -297.13 \\ 0 & 0 & -1 & 0 \end{pmatrix}$
Camera pose	$V = \begin{pmatrix} 1 & 0 & 0 & 0 \\ 0 & 0 & 1 & 0 \\ 0 & -1 & 0 & -520 \\ 0 & 0 & 0 & 1 \end{pmatrix}$	$(R\|t) = \begin{pmatrix} 1 & 7.67e^{-7} & -1.71e^{-7} & 4.56e^{-4} \\ -1.71e^{-7} & -4.57e^{-6} & -1 & -1.58e^{-3} \\ -7.67e^{-7} & 1 & -4.57e^{-6} & 519.98 \end{pmatrix}$	$V_{\text{cal.}} = \begin{pmatrix} 1 & 7.67e^{-7} & -1.71e^{-7} & 4.56e^{-4} \\ 1.71e^{-7} & 4.57e^{-6} & 1 & 1.58e^{-3} \\ 7.67e^{-7} & -1 & 4.57e^{-6} & -519.98 \\ 0 & 0 & 0 & 1 \end{pmatrix}$
Distortion	$k_1 = 0 \quad k_4 = 0$ $k_2 = 0 \quad k_5 = 0$ $k_3 = 0 \quad k_6 = 0$ $p_1 = 0 \quad s_1 = 0$ $p_2 = 0 \quad s_2 = 0$ $\tau_x = 0 \quad s_3 = 0$ $\tau_y = 0 \quad s_4 = 0$	$k_1 = 0 \quad k_4 = 0$ $k_2 = 0 \quad k_5 = 0$ $k_3 = 0 \quad k_6 = 0$ $p_1 = 0 \quad s_1 = 0$ $p_2 = 0 \quad s_2 = 0$ $\tau_x = 0 \quad s_3 = 0$ $\tau_y = 0 \quad s_4 = 0$	
Pattern pose	$M = \begin{pmatrix} 1 & 0 & 0 & 0 \\ 0 & 1 & 0 & -7.5 \\ 0 & 0 & 1 & 0 \\ 0 & 0 & 0 & 1 \end{pmatrix}$	$M_{\text{cal.}} = \begin{pmatrix} 1 & 0 & 0 & 0 \\ 0 & 1 & 0 & -7.5 \\ 0 & 0 & 1 & 0 \\ 0 & 0 & 0 & 1 \end{pmatrix}$ Intrinsic calibration: Repr. error $= 0.043374$ px, views $= 31$ Extrinsic calibration: Repr. error $= 0.041648$ px, views $= 31$	

Table A.4 – Test #3: Principal point c_x, c_y not in image center (modified projection matrix P).

126

	Ground truth	Calibrated camera model	Validation
Parameters	width $= 1584$, height $= 2376$ $\mathrm{fov}_y = 18.5°$ $c_x = 791.5$, $c_y = 1187.5$	$f_x \overset{!}{=} f_y,$ $p_1 \overset{!}{=} 0$, $p_2 \overset{!}{=} 0$, $k_1 \overset{!}{=} 0$, $k_2 \overset{!}{=} 0$, $k_3 \overset{!}{=} 0,$ Estimate misalignment: False	
Projection	$P = \begin{pmatrix} 9.2103 & 0 & 0 & 0 \\ 0 & 6.1402 & 0 & 0 \\ 0 & 0 & -2.5 & -1050 \\ 0 & 0 & -1 & 0 \end{pmatrix}$	$K = \begin{pmatrix} 7294.4 & 0 & 790.35 \\ 0 & 7294.4 & 1187.6 \\ 0 & 0 & 1 \end{pmatrix}$	$P_{\text{cal.}} = \begin{pmatrix} 9.2101 & 0 & 1.45e^{-3} & 0 \\ 0 & 6.14 & 5.83e^{-5} & 0 \\ 0 & 0 & -1.2857 & -297.13 \\ 0 & 0 & -1 & 0 \end{pmatrix}$
Camera pose	$V = \begin{pmatrix} 1 & 0 & 0 & 7 \\ 0 & 0 & 1 & -4 \\ 0 & -1 & 0 & -520 \\ 0 & 0 & 0 & 1 \end{pmatrix}$	$(R\|t) = \begin{pmatrix} 1 & 1.68e^{-4} & -3.17e^{-6} & 7.0817 \\ -3.17e^{-6} & 3.10e^{-7} & -1 & 3.9949 \\ -1.68e^{-4} & 1 & 3.10e^{-7} & 519.98 \end{pmatrix}$	$V_{\text{cal.}} = \begin{pmatrix} 1 & 1.68e^{-4} & -3.17e^{-6} & 7.0817 \\ 3.17e^{-6} & -3.10e^{-7} & 1 & -3.9949 \\ 1.68e^{-4} & -1 & -3.10e^{-7} & -519.98 \\ 0 & 0 & 0 & 1 \end{pmatrix}$
Distortion	$k_1 = 0 \quad k_4 = 0$ $k_2 = 0 \quad k_5 = 0$ $k_3 = 0 \quad k_6 = 0$ $p_1 = 0 \quad s_1 = 0$ $p_2 = 0 \quad s_2 = 0$ $\tau_x = 0 \quad s_3 = 0$ $\tau_y = 0 \quad s_4 = 0$	$k_1 = 0 \quad k_4 = 0$ $k_2 = 0 \quad k_5 = 0$ $k_3 = 0 \quad k_6 = 0$ $p_1 = 0 \quad s_1 = 0$ $p_2 = 0 \quad s_2 = 0$ $\tau_x = 0 \quad s_3 = 0$ $\tau_y = 0 \quad s_4 = 0$	
Pattern pose	$M = \begin{pmatrix} 1 & 0 & 0 & 0 \\ 0 & 1 & 0 & -7.5 \\ 0 & 0 & 1 & 0 \\ 0 & 0 & 0 & 1 \end{pmatrix}$	$M_{\text{cal.}} = \begin{pmatrix} 1 & 0 & 0 & 0 \\ 0 & 1 & 0 & -7.5 \\ 0 & 0 & 1 & 0 \\ 0 & 0 & 0 & 1 \end{pmatrix}$ Intrinsic calibration: Repr. error $= 0.043695\,$px, views $= 31$ Extrinsic calibration: Repr. error $= 0.041764\,$px, views $= 31$	

Table A.5 – Test #4: Camera translation $\Delta x = 7$, $\Delta y = -4$ (modified view matrix V).

	Ground truth	Calibrated camera model	Validation
Parameters	$\text{width} = 1584,\ \text{height} = 2376$	$f_x \overset{!}{=} f_y,$	
	$\text{fov}_y = 18.5°$	$p_1 \overset{!}{=} 0,\ p_2 \overset{!}{=} 0,\ k_1 \overset{!}{=} 0,\ k_2 \overset{!}{=} 0,\ k_3 \overset{!}{=} 0,$	
	$c_x = 791.5,\ c_y = 1187.5$	Estimate misalignment: False	
Projection	$P = \begin{pmatrix} 9.2103 & 0 & 0 & 0 \\ 0 & 6.1402 & 0 & 0 \\ 0 & 0 & -2.5 & -1050 \\ 0 & 0 & -1 & 0 \end{pmatrix}$	$K = \begin{pmatrix} 7295.9 & 0 & 791.54 \\ 0 & 7295.9 & 1188.3 \\ 0 & 0 & 1 \end{pmatrix}$	$P_{\text{cal.}} = \begin{pmatrix} 9.212 & 0 & -4.54e^{-5} & 0 \\ 0 & 6.1414 & 6.32e^{-4} & 0 \\ 0 & 0 & -1.2857 & -297.2 \\ 0 & 0 & -1 & 0 \end{pmatrix}$
Camera pose	$V = \begin{pmatrix} 1 & 0 & 0 & 0 \\ 0 & 0.087156 & 0.99619 & 0 \\ 0 & -0.99619 & 0.087156 & -520 \\ 0 & 0 & 0 & 1 \end{pmatrix}$	$(R\|t) = \begin{pmatrix} 1 & -9.55e^{-6} & -1.93e^{-6} & -2.63e^{-3} \\ -2.76e^{-6} & -0.087234 & -0.99619 & -0.053306 \\ 9.35e^{-6} & 0.99619 & -0.087234 & 520.1 \end{pmatrix}$	$V_{\text{cal.}} = \begin{pmatrix} 1 & -9.55e^{-6} & -1.93e^{-6} & -2.63e^{-3} \\ 2.76e^{-6} & 0.087234 & 0.99619 & 0.053306 \\ -9.35e^{-6} & -0.99619 & 0.087234 & -520.1 \\ 0 & 0 & 0 & 1 \end{pmatrix}$
Distortion	$k_1 = 0 \quad k_4 = 0$ $k_2 = 0 \quad k_5 = 0$ $k_3 = 0 \quad k_6 = 0$ $p_1 = 0 \quad s_1 = 0$ $p_2 = 0 \quad s_2 = 0$ $\tau_x = 0 \quad s_3 = 0$ $\tau_y = 0 \quad s_4 = 0$	$k_1 = 0 \quad k_4 = 0$ $k_2 = 0 \quad k_5 = 0$ $k_3 = 0 \quad k_6 = 0$ $p_1 = 0 \quad s_1 = 0$ $p_2 = 0 \quad s_2 = 0$ $\tau_x = 0 \quad s_3 = 0$ $\tau_y = 0 \quad s_4 = 0$	
Pattern pose	$M = \begin{pmatrix} 1 & 0 & 0 & 0 \\ 0 & 1 & 0 & -7.5 \\ 0 & 0 & 1 & 0 \\ 0 & 0 & 0 & 1 \end{pmatrix}$	$M_{\text{cal.}} = \begin{pmatrix} 1 & 0 & 0 & 0 \\ 0 & 1 & 0 & -7.5 \\ 0 & 0 & 1 & 0 \\ 0 & 0 & 0 & 1 \end{pmatrix}$ Intrinsic calibration: Repr. error = 0.049413 px, views = 31 Extrinsic calibration: Repr. error = 0.048661 px, views = 31	

Table **A.6** – Test #5: Camera rotated 5° about x-axis (modified view matrix V).

	Ground truth	Calibrated camera model	Validation
Parameters	width $= 1584$, height $= 2376$ fov$_y = 18.5°$ $c_x = 791.5$, $c_y = 1187.5$	$f_x \overset{!}{=} f_y,$ $p_1 \overset{!}{=} 0,\ p_2 \overset{!}{=} 0,\ k_1 \overset{!}{=} 0,\ k_2 \overset{!}{=} 0,\ k_3 \overset{!}{=} 0,$ Estimate misalignment: False	
Projection	$P = \begin{pmatrix} 9.2103 & 0 & 0 & 0 \\ 0 & 6.1402 & 0 & 0 \\ 0 & 0 & -2.5 & -1050 \\ 0 & 0 & -1 & 0 \end{pmatrix}$	$K = \begin{pmatrix} 7294.6 & 0 & 791.48 \\ 0 & 7294.6 & 1187.5 \\ 0 & 0 & 1 \end{pmatrix}$	$P_{\text{cal.}} = \begin{pmatrix} 9.2104 & 0 & 2.31\mathrm{e}^{-5} & 0 \\ 0 & 6.1402 & -7.93\mathrm{e}^{-6} & 0 \\ 0 & 0 & -1.2857 & -297.14 \\ 0 & 0 & -1 & 0 \end{pmatrix}$
Camera pose	$V = \begin{pmatrix} 0.99619 & 0 & 0.087156 & 0 \\ -0.087156 & 0 & 0.99619 & 0 \\ 0 & -1 & 0 & -520 \\ 0 & 0 & 0 & 1 \end{pmatrix}$	$(R\|t) = \begin{pmatrix} 0.99619 & 8.76\mathrm{e}^{-6} & 0.087156 & 1.39\mathrm{e}^{-3} \\ 0.087156 & 1.29\mathrm{e}^{-5} & -0.99619 & 7.07\mathrm{e}^{-4} \\ -9.85\mathrm{e}^{-6} & 1 & 1.21\mathrm{e}^{-5} & 520 \end{pmatrix}$	$V_{\text{cal.}} = \begin{pmatrix} 0.99619 & 8.76\mathrm{e}^{-6} & 0.087156 & 1.39\mathrm{e}^{-3} \\ -0.087156 & -1.29\mathrm{e}^{-5} & 0.99619 & -7.07\mathrm{e}^{-4} \\ 9.85\mathrm{e}^{-6} & -1 & -1.21\mathrm{e}^{-5} & -520 \\ 0 & 0 & 0 & 1 \end{pmatrix}$
Distortion	$k_1 = 0 \quad k_4 = 0$ $k_2 = 0 \quad k_5 = 0$ $k_3 = 0 \quad k_6 = 0$ $p_1 = 0 \quad s_1 = 0$ $p_2 = 0 \quad s_2 = 0$ $\tau_x = 0 \quad s_3 = 0$ $\tau_y = 0 \quad s_4 = 0$	$k_1 = 0 \quad k_4 = 0$ $k_2 = 0 \quad k_5 = 0$ $k_3 = 0 \quad k_6 = 0$ $p_1 = 0 \quad s_1 = 0$ $p_2 = 0 \quad s_2 = 0$ $\tau_x = 0 \quad s_3 = 0$ $\tau_y = 0 \quad s_4 = 0$	
Pattern pose	$M = \begin{pmatrix} 1 & 0 & 0 & 0 \\ 0 & 1 & 0 & -7.5 \\ 0 & 0 & 1 & 0 \\ 0 & 0 & 0 & 1 \end{pmatrix}$	$M_{\text{cal.}} = \begin{pmatrix} 1 & 0 & 0 & 0 \\ 0 & 1 & 0 & -7.5 \\ 0 & 0 & 1 & 0 \\ 0 & 0 & 0 & 1 \end{pmatrix}$ Intrinsic calibration: Repr. error $= 0.024921$ px, views $= 31$ Extrinsic calibration: Repr. error $= 0.022688$ px, views $= 31$	

Table A.7 – Test #6: Camera rotated 5° about y-axis (modified view matrix V).

	Ground truth	Calibrated camera model	Validation
Parameters	$\text{width} = 1584$, $\text{height} = 2376$ $\text{fov}_y = 18.5°$ $c_x = 791.5$, $c_y = 1187.5$	$f_x \overset{!}{=} f_y,$ $p_1 \overset{!}{=} 0,\ p_2 \overset{!}{=} 0,\ k_1 \overset{!}{=} 0,\ k_2 \overset{!}{=} 0,\ k_3 \overset{!}{=} 0,$ Estimate misalignment: False	
Projection	$P = \begin{pmatrix} 9.2103 & 0 & 0 & 0 \\ 0 & 6.1402 & 0 & 0 \\ 0 & 0 & -2.5 & -1050 \\ 0 & 0 & -1 & 0 \end{pmatrix}$	$K = \begin{pmatrix} 7294.3 & 0 & 791.37 \\ 0 & 7294.3 & 1187.5 \\ 0 & 0 & 1 \end{pmatrix}$	$P_{\text{cal.}} = \begin{pmatrix} 9.21 & 0 & 1.67\text{e}^{-4} & 0 \\ 0 & 6.14 & 9.54\text{e}^{-6} & 0 \\ 0 & 0 & -1.2857 & -297.13 \\ 0 & 0 & -1 & 0 \end{pmatrix}$
Camera pose	$V = \begin{pmatrix} 0.99619 & -0.087156 & 0 & 0 \\ 0 & 0 & 1 & 0 \\ -0.087156 & -0.99619 & 0 & -520 \\ 0 & 0 & 0 & 1 \end{pmatrix}$	$(R\mid t) = \begin{pmatrix} 0.9962 & -0.087139 & -4.80\text{e}^{-8} & 9.39\text{e}^{-3} \\ -1.89\text{e}^{-7} & -1.61\text{e}^{-6} & -1 & -9.19\text{e}^{-4} \\ 0.087139 & 0.9962 & -1.62\text{e}^{-6} & 519.98 \end{pmatrix}$	$V_{\text{cal.}} = \begin{pmatrix} 0.9962 & -0.087139 & -4.80\text{e}^{-8} & 9.39\text{e}^{-3} \\ 1.89\text{e}^{-7} & 1.61\text{e}^{-6} & 1 & 9.19\text{e}^{-4} \\ -0.087139 & -0.9962 & 1.62\text{e}^{-6} & -519.98 \\ 0 & 0 & 0 & 1 \end{pmatrix}$
Distortion	$k_1 = 0 \quad k_4 = 0$ $k_2 = 0 \quad k_5 = 0$ $k_3 = 0 \quad k_6 = 0$ $p_1 = 0 \quad s_1 = 0$ $p_2 = 0 \quad s_2 = 0$ $\tau_x = 0 \quad s_3 = 0$ $\tau_y = 0 \quad s_4 = 0$	$k_1 = 0 \quad k_4 = 0$ $k_2 = 0 \quad k_5 = 0$ $k_3 = 0 \quad k_6 = 0$ $p_1 = 0 \quad s_1 = 0$ $p_2 = 0 \quad s_2 = 0$ $\tau_x = 0 \quad s_3 = 0$ $\tau_y = 0 \quad s_4 = 0$	
Pattern pose	$M = \begin{pmatrix} 1 & 0 & 0 & 0 \\ 0 & 1 & 0 & -7.5 \\ 0 & 0 & 1 & 0 \\ 0 & 0 & 0 & 1 \end{pmatrix}$	$M_{\text{cal.}} = \begin{pmatrix} 1 & 0 & 0 & 0 \\ 0 & 1 & 0 & -7.5 \\ 0 & 0 & 1 & 0 \\ 0 & 0 & 0 & 1 \end{pmatrix}$ Intrinsic calibration: Repr. error $= 0.043198\,\text{px}$, views $= 31$ Extrinsic calibration: Repr. error $= 0.041584\,\text{px}$, views $= 31$	

Table A.8 – Test #7: Camera rotated 5° about z-axis (modified view matrix V).

	Ground truth	Calibrated camera model	Validation
Parameters	width $= 1584$, height $= 2376$	$f_x \overset{!}{=} f_y,$	
	$\text{fov}_y = 18.5°$	$p_1 \overset{!}{=} 0,\ p_2 \overset{!}{=} 0,\ k_1 \overset{!}{=} 0,\ k_2 \overset{!}{=} 0,\ k_3 \overset{!}{=} 0,$	
	$c_x = 791.5,\ c_y = 1187.5$	Estimate misalignment: False	
Projection	$P = \begin{pmatrix} 9.2103 & 0 & 0 & 0 \\ 0 & 6.1402 & 0 & 0 \\ 0 & 0 & -2.5 & -1050 \\ 0 & 0 & -1 & 0 \end{pmatrix}$	$K = \begin{pmatrix} 7294.7 & 0 & 791.69 \\ 0 & 7294.7 & 1187.4 \\ 0 & 0 & 1 \end{pmatrix}$	$P_{\text{cal.}} = \begin{pmatrix} 9.2105 & 0 & -2.42e^{-4} & 0 \\ 0 & 6.1403 & -8.65e^{-5} & 0 \\ 0 & 0 & -1.2857 & -297.15 \\ 0 & 0 & -1 & 0 \end{pmatrix}$
Camera pose	$V = \begin{pmatrix} 0.99624 & -0.07766 & -0.03836 & 0 \\ 0.039112 & 8.18e^{-3} & 0.9992 & 0 \\ -0.077284 & -0.99695 & 0.011187 & -520 \\ 0 & 0 & 0 & 1 \end{pmatrix}$	$(R|t) = \begin{pmatrix} 0.99624 & -0.077693 & -0.038358 & -0.013652 \\ -0.039109 & -8.16e^{-3} & -0.9992 & 7.25e^{-3} \\ 0.077318 & 0.99694 & -0.011171 & 520.01 \end{pmatrix}$	$V_{\text{cal.}} = \begin{pmatrix} 0.99624 & -0.077693 & -0.038358 & -0.013652 \\ 0.039109 & 8.16e^{-3} & 0.9992 & -7.25e^{-3} \\ -0.077318 & -0.99694 & 0.011171 & -520.01 \\ 0 & 0 & 0 & 1 \end{pmatrix}$
Distortion	$k_1 = 0 \quad k_4 = 0$ $k_2 = 0 \quad k_5 = 0$ $k_3 = 0 \quad k_6 = 0$ $p_1 = 0 \quad s_1 = 0$ $p_2 = 0 \quad s_2 = 0$ $\tau_x = 0 \quad s_3 = 0$ $\tau_y = 0 \quad s_4 = 0$	$k_1 = 0 \quad k_4 = 0$ $k_2 = 0 \quad k_5 = 0$ $k_3 = 0 \quad k_6 = 0$ $p_1 = 0 \quad s_1 = 0$ $p_2 = 0 \quad s_2 = 0$ $\tau_x = 0 \quad s_3 = 0$ $\tau_y = 0 \quad s_4 = 0$	
Pattern pose	$M = \begin{pmatrix} 1 & 0 & 0 & 0 \\ 0 & 1 & 0 & -7.5 \\ 0 & 0 & 1 & 0 \\ 0 & 0 & 0 & 1 \end{pmatrix}$	$M_{\text{cal.}} = \begin{pmatrix} 1 & 0 & 0 & 0 \\ 0 & 1 & 0 & -7.5 \\ 0 & 0 & 1 & 0 \\ 0 & 0 & 0 & 1 \end{pmatrix}$ Intrinsic calibration: Repr. error $= 0.026454$ px, views $= 31$ Extrinsic calibration: Repr. error $= 0.023684$ px, views $= 31$	

Table A.9 – Test #8: Camera rotated 5° about skew axis $(1, -4, 8)$ (modified view matrix V).

	Ground truth	Calibrated camera model	Validation
Parameters	width $= 1584$, height $= 2376$ fov$_y = 18.5°$ $c_x = 791.5$, $c_y = 1187.5$	$f_x \overset{!}{=} f_y,$ $p_1 \overset{!}{=} 0$, $p_2 \overset{!}{=} 0$, $k_1 \overset{!}{=} 0$, $k_2 \overset{!}{=} 0$, $k_3 \overset{!}{=} 0,$ Estimate misalignment: True	
Projection	$P = \begin{pmatrix} 9.2103 & 0 & 0 & 0 \\ 0 & 6.1402 & 0 & 0 \\ 0 & 0 & -2.5 & -1050 \\ 0 & 0 & -1 & 0 \end{pmatrix}$	$K = \begin{pmatrix} 7294.2 & 0 & 792.24 \\ 0 & 7294.2 & 1187.5 \\ 0 & 0 & 1 \end{pmatrix}$	$P_{\text{cal.}} = \begin{pmatrix} 9.2098 & 0 & -9.36e^{-4} & 0 \\ 0 & 6.1399 & -2.68e^{-5} & 0 \\ 0 & 0 & -1.2857 & -297.13 \\ 0 & 0 & -1 & 0 \end{pmatrix}$
Camera pose	$V = \begin{pmatrix} 1 & 0 & 0 & 0 \\ 0 & 0 & 1 & 0 \\ 0 & -1 & 0 & -520 \\ 0 & 0 & 0 & 1 \end{pmatrix}$	$(R|t) = \begin{pmatrix} 1 & -1.05e^{-4} & -3.78e^{-6} & -0.052876 \\ -3.78e^{-6} & 1.65e^{-5} & -1 & 2.42e^{-3} \\ 1.05e^{-4} & 1 & 1.65e^{-5} & 519.97 \end{pmatrix}$	$V_{\text{cal.}} = \begin{pmatrix} 1 & -1.05e^{-4} & -3.78e^{-6} & -0.052876 \\ 3.78e^{-6} & -1.65e^{-5} & 1 & -2.42e^{-3} \\ -1.05e^{-4} & -1 & -1.65e^{-5} & -519.97 \\ 0 & 0 & 0 & 1 \end{pmatrix}$
Distortion	$k_1 = 0 \quad k_4 = 0$ $k_2 = 0 \quad k_5 = 0$ $k_3 = 0 \quad k_6 = 0$ $p_1 = 0 \quad s_1 = 0$ $p_2 = 0 \quad s_2 = 0$ $\tau_x = 0 \quad s_3 = 0$ $\tau_y = 0 \quad s_4 = 0$	$k_1 = 0 \quad k_4 = 0$ $k_2 = 0 \quad k_5 = 0$ $k_3 = 0 \quad k_6 = 0$ $p_1 = 0 \quad s_1 = 0$ $p_2 = 0 \quad s_2 = 0$ $\tau_x = 0 \quad s_3 = 0$ $\tau_y = 0 \quad s_4 = 0$	
Pattern pose	$M = \begin{pmatrix} 1 & 0 & 0 & -3 \\ 0 & 1 & 0 & -8 \\ 0 & 0 & 1 & 0 \\ 0 & 0 & 0 & 1 \end{pmatrix}$	$M_{\text{cal.}} = \begin{pmatrix} 1 & 7.81e^{-11} & 6.63e^{-6} & -3 \\ 0 & 1 & -1.18e^{-5} & -8.0008 \\ -6.63e^{-6} & 1.18e^{-5} & 1 & 0 \\ 0 & 0 & 0 & 1 \end{pmatrix}$ Intrinsic calibration: Repr. error $= 0.043799$ px, views $= 31$ Extrinsic calibration: Repr. error $= 0.042506$ px, views $= 31$	

Table A.10 – Test #9: Pattern misalignment 1: translation $\Delta x = -3$, $\Delta y = -0.5$ (modified matrix M).

	Ground truth	Calibrated camera model	Validation
Parameters	width = 1584, height = 2376 $\text{fov}_y = 18.5°$ $c_x = 791.5, c_y = 1187.5$	$f_x \overset{!}{=} f_y,$ $p_1 \overset{!}{=} 0, p_2 \overset{!}{=} 0, k_1 \overset{!}{=} 0, k_2 \overset{!}{=} 0, k_3 \overset{!}{=} 0,$ Estimate misalignment: True	
Projection	$P = \begin{pmatrix} 9.2103 & 0 & 0 & 0 \\ 0 & 6.1402 & 0 & 0 \\ 0 & 0 & -2.5 & -1050 \\ 0 & 0 & -1 & 0 \end{pmatrix}$	$K = \begin{pmatrix} 7294.7 & 0 & 791.54 \\ 0 & 7294.7 & 1187.5 \\ 0 & 0 & 1 \end{pmatrix}$	$P_{\text{cal.}} = \begin{pmatrix} 9.2105 & 0 & -5.64e^{-5} & 0 \\ 0 & 6.1403 & -1.12e^{-5} & 0 \\ 0 & 0 & -1.2857 & -297.15 \\ 0 & 0 & -1 & 0 \end{pmatrix}$
Camera pose	$V = \begin{pmatrix} 1 & 0 & 0 & 0 \\ 0 & 0 & 1 & 0 \\ 0 & -1 & 0 & -520 \\ 0 & 0 & 0 & 1 \end{pmatrix}$	$(R\|t) = \begin{pmatrix} 1 & -7.66e^{-6} & -5.78e^{-8} & -3.32e^{-3} \\ -5.78e^{-8} & 7.35e^{-6} & -1 & -0.65262 \\ 7.66e^{-6} & 1 & 7.35e^{-6} & 520.01 \end{pmatrix}$	$V_{\text{cal.}} = \begin{pmatrix} 1 & -7.66e^{-6} & -5.78e^{-8} & -3.32e^{-3} \\ 5.78e^{-8} & -7.35e^{-6} & 1 & 0.65262 \\ -7.66e^{-6} & -1 & -7.35e^{-6} & -520.01 \\ 0 & 0 & 0 & 1 \end{pmatrix}$
Distortion	$k_1 = 0 \quad k_4 = 0$ $k_2 = 0 \quad k_5 = 0$ $k_3 = 0 \quad k_6 = 0$ $p_1 = 0 \quad s_1 = 0$ $p_2 = 0 \quad s_2 = 0$ $\tau_x = 0 \quad s_3 = 0$ $\tau_y = 0 \quad s_4 = 0$	$k_1 = 0 \quad k_4 = 0$ $k_2 = 0 \quad k_5 = 0$ $k_3 = 0 \quad k_6 = 0$ $p_1 = 0 \quad s_1 = 0$ $p_2 = 0 \quad s_2 = 0$ $\tau_x = 0 \quad s_3 = 0$ $\tau_y = 0 \quad s_4 = 0$	
Pattern pose	$M = \begin{pmatrix} 1 & 0 & 0 & 0 \\ 0 & 0.99619 & 0.087156 & -7.5 \\ 0 & -0.087156 & 0.99619 & 0 \\ 0 & 0 & 0 & 1 \end{pmatrix}$	$M_{\text{cal.}} = \begin{pmatrix} 1 & -2.29e^{-8} & 2.62e^{-7} & 1.79e^{-4} \\ 0 & 0.99619 & 0.087162 & -7.4713 \\ -2.63e^{-7} & -0.087162 & 0.99619 & 0 \\ 0 & 0 & 0 & 1 \end{pmatrix}$ Intrinsic calibration: Repr. error = 0.03438 px, views = 31 Extrinsic calibration: Repr. error = 0.02737 px, views = 31	

Table A.11 – Test #10: Pattern misalignment 2: rotation −5° about x-axis (modified matrix M).

	Ground truth	Calibrated camera model	Validation
Parameters	width $= 1584$, height $= 2376$ fov$_y = 18.5°$ $c_x = 791.5$, $c_y = 1187.5$	$f_x \overset{!}{=} f_y,$ $p_1 \overset{!}{=} 0$, $p_2 \overset{!}{=} 0$, $k_1 \overset{!}{=} 0$, $k_2 \overset{!}{=} 0$, $k_3 \overset{!}{=} 0,$ Estimate misalignment: True	
Projection	$P = \begin{pmatrix} 9.2103 & 0 & 0 & 0 \\ 0 & 6.1402 & 0 & 0 \\ 0 & 0 & -2.5 & -1050 \\ 0 & 0 & -1 & 0 \end{pmatrix}$	$K = \begin{pmatrix} 7294.3 & 0 & 791.52 \\ 0 & 7294.3 & 1187.5 \\ 0 & 0 & 1 \end{pmatrix}$	$P_{\text{cal.}} = \begin{pmatrix} 9.21 & 0 & -2.86\text{e}^{-5} & 0 \\ 0 & 6.14 & -4.04\text{e}^{-5} & 0 \\ 0 & 0 & -1.2857 & -297.13 \\ 0 & 0 & -1 & 0 \end{pmatrix}$
Camera pose	$V = \begin{pmatrix} 1 & 0 & 0 & 0 \\ 0 & 0 & 1 & 0 \\ 0 & -1 & 0 & -520 \\ 0 & 0 & 0 & 1 \end{pmatrix}$	$(R\lvert t) = \begin{pmatrix} 1 & 8.39\text{e}^{-6} & -6.03\text{e}^{-6} & -1.58\text{e}^{-3} \\ -6.03\text{e}^{-6} & 1.86\text{e}^{-5} & -1 & 3.50\text{e}^{-3} \\ -8.39\text{e}^{-6} & 1 & 1.86\text{e}^{-5} & 519.98 \end{pmatrix}$	$V_{\text{cal.}} = \begin{pmatrix} 1 & 8.39\text{e}^{-6} & -6.03\text{e}^{-6} & -1.58\text{e}^{-3} \\ 6.03\text{e}^{-6} & -1.86\text{e}^{-5} & 1 & -3.50\text{e}^{-3} \\ 8.39\text{e}^{-6} & -1 & -1.86\text{e}^{-5} & -519.98 \\ 0 & 0 & 0 & 1 \end{pmatrix}$
Distortion	$k_1 = 0 \quad k_4 = 0$ $k_2 = 0 \quad k_5 = 0$ $k_3 = 0 \quad k_6 = 0$ $p_1 = 0 \quad s_1 = 0$ $p_2 = 0 \quad s_2 = 0$ $\tau_x = 0 \quad s_3 = 0$ $\tau_y = 0 \quad s_4 = 0$	$k_1 = 0 \quad k_4 = 0$ $k_2 = 0 \quad k_5 = 0$ $k_3 = 0 \quad k_6 = 0$ $p_1 = 0 \quad s_1 = 0$ $p_2 = 0 \quad s_2 = 0$ $\tau_x = 0 \quad s_3 = 0$ $\tau_y = 0 \quad s_4 = 0$	
Pattern pose	$M = \begin{pmatrix} 0.99619 & 0 & 0.087156 & 0 \\ 0 & 1 & 0 & -7.5 \\ -0.087156 & 0 & 0.99619 & 0 \\ 0 & 0 & 0 & 1 \end{pmatrix}$	$M_{\text{cal.}} = \begin{pmatrix} 0.99619 & -2.10\text{e}^{-8} & 0.087158 & 5.91\text{e}^{-5} \\ 0 & 1 & 2.41\text{e}^{-7} & -7.4999 \\ -0.087158 & -2.40\text{e}^{-7} & 0.99619 & 0 \\ 0 & 0 & 0 & 1 \end{pmatrix}$ Intrinsic calibration: Repr. error $= 0.032948$ px, views $= 31$ Extrinsic calibration: Repr. error $= 0.026948$ px, views $= 31$	

Table A.12 – Test #11: Pattern misalignment 3: rotation 5° about y-axis (modified matrix M).

	Ground truth	Calibrated camera model	Validation
Parameters	width $= 1584$, height $= 2376$ $\text{fov}_y = 18.5°$ $c_x = 791.5$, $c_y = 1187.5$	$f_x \overset{!}{=} f_y,$ $p_1 \overset{!}{=} 0$, $p_2 \overset{!}{=} 0$, $k_1 \overset{!}{=} 0$, $k_2 \overset{!}{=} 0$, $k_3 \overset{!}{=} 0,$ Estimate misalignment: True	
Projection	$P = \begin{pmatrix} 9.2103 & 0 & 0 & 0 \\ 0 & 6.1402 & 0 & 0 \\ 0 & 0 & -2.5 & -1050 \\ 0 & 0 & -1 & 0 \end{pmatrix}$	$K = \begin{pmatrix} 7295.1 & 0 & 791.26 \\ 0 & 7295.1 & 1187.8 \\ 0 & 0 & 1 \end{pmatrix}$	$P_{\text{cal.}} = \begin{pmatrix} 9.211 & 0 & 3.04\mathrm{e}^{-4} & 0 \\ 0 & 6.1406 & 2.24\mathrm{e}^{-4} & 0 \\ 0 & 0 & -1.2857 & -297.16 \\ 0 & 0 & -1 & 0 \end{pmatrix}$
Camera pose	$V = \begin{pmatrix} 1 & 0 & 0 & 0 \\ 0 & 0 & 1 & 0 \\ 0 & -1 & 0 & -520 \\ 0 & 0 & 0 & 1 \end{pmatrix}$	$(R\|t) = \begin{pmatrix} 1 & 4.99\mathrm{e}^{-5} & 4.95\mathrm{e}^{-6} & 0.017536 \\ 4.95\mathrm{e}^{-6} & -2.52\mathrm{e}^{-5} & -1 & -0.67 \\ -4.99\mathrm{e}^{-5} & 1 & -2.52\mathrm{e}^{-5} & 520.03 \end{pmatrix}$	$V_{\text{cal.}} = \begin{pmatrix} 1 & 4.99\mathrm{e}^{-5} & 4.95\mathrm{e}^{-6} & 0.017536 \\ -4.95\mathrm{e}^{-6} & 2.52\mathrm{e}^{-5} & 1 & 0.67 \\ 4.99\mathrm{e}^{-5} & -1 & 2.52\mathrm{e}^{-5} & -520.03 \\ 0 & 0 & 0 & 1 \end{pmatrix}$
Distortion	$k_1 = 0 \quad k_4 = 0$ $k_2 = 0 \quad k_5 = 0$ $k_3 = 0 \quad k_6 = 0$ $p_1 = 0 \quad s_1 = 0$ $p_2 = 0 \quad s_2 = 0$ $\tau_x = 0 \quad s_3 = 0$ $\tau_y = 0 \quad s_4 = 0$	$k_1 = 0 \quad k_4 = 0$ $k_2 = 0 \quad k_5 = 0$ $k_3 = 0 \quad k_6 = 0$ $p_1 = 0 \quad s_1 = 0$ $p_2 = 0 \quad s_2 = 0$ $\tau_x = 0 \quad s_3 = 0$ $\tau_y = 0 \quad s_4 = 0$	
Pattern pose	$M = \begin{pmatrix} 0.99619 & -7.60\mathrm{e}^{-3} & 0.086824 & -3 \\ 0 & 0.99619 & 0.087156 & -7 \\ -0.087156 & -0.086824 & 0.9924 & 0 \\ 0 & 0 & 0 & 1 \end{pmatrix}$	$M_{\text{cal.}} = \begin{pmatrix} 0.9962 & -7.60\mathrm{e}^{-3} & 0.086819 & -2.9433 \\ 0 & 0.9962 & 0.08715 & -6.9715 \\ -0.08715 & -0.086819 & 0.9924 & 0 \\ 0 & 0 & 0 & 1 \end{pmatrix}$ Intrinsic calibration: Repr. error $= 0.036955\,\text{px}$, views $= 31$ Extrinsic calibration: Repr. error $= 0.026317\,\text{px}$, views $= 31$	

Table A.13 – Test #12: Pattern misalignment 4: combination of tests #9, #10 and #11.

	Ground truth	Calibrated camera model	Validation
Parameters	$\text{width} = 1584,\ \text{height} = 2376$ $\text{fov}_y = 18.5°$ $c_x = 791.5,\ c_y = 1187.5$	$f_x \overset{!}{=} f_y,$ $p_1 \overset{!}{=} 0,\ p_2 \overset{!}{=} 0,\ k_1 \overset{!}{=} 0,\ k_2 \overset{!}{=} 0,\ k_3 \overset{!}{=} 0,$ Estimate misalignment: False	
Projection	$P = \begin{pmatrix} 9.2103 & 0 & 0 & 0 \\ 0 & 6.1402 & 0 & 0 \\ 0 & 0 & -2.5 & -1050 \\ 0 & 0 & -1 & 0 \end{pmatrix}$	$K = \begin{pmatrix} 7295.1 & 0 & 791.26 \\ 0 & 7295.1 & 1187.8 \\ 0 & 0 & 1 \end{pmatrix}$	$P_{\text{cal.}} = \begin{pmatrix} 9.211 & 0 & 3.04\mathrm{e}^{-4} & 0 \\ 0 & 6.1406 & 2.24\mathrm{e}^{-4} & 0 \\ 0 & 0 & -1.2857 & -294.02 \\ 0 & 0 & -1 & 0 \end{pmatrix}$
Camera pose	$V = \begin{pmatrix} 1 & 0 & 0 & 0 \\ 0 & 0 & 1 & 0 \\ 0 & -1 & 0 & -520 \\ 0 & 0 & 0 & 1 \end{pmatrix}$	$(R\|t) = \begin{pmatrix} 0.99704 & 0.019392 & 0.074376 & -1.6213 \\ 0.074385 & 2.60\mathrm{e}^{-4} & -0.99723 & -0.96742 \\ -0.019358 & 0.99981 & -1.18\mathrm{e}^{-3} & 514.54 \end{pmatrix}$	$V_{\text{cal.}} = \begin{pmatrix} 0.99704 & 0.019392 & 0.074376 & -1.6213 \\ -0.074385 & -2.60\mathrm{e}^{-4} & 0.99723 & 0.96742 \\ 0.019358 & -0.99981 & 1.18\mathrm{e}^{-3} & -514.54 \\ 0 & 0 & 0 & 1 \end{pmatrix}$
Distortion	$\begin{aligned} k_1 &= 0 & k_4 &= 0 \\ k_2 &= 0 & k_5 &= 0 \\ k_3 &= 0 & k_6 &= 0 \\ p_1 &= 0 & s_1 &= 0 \\ p_2 &= 0 & s_2 &= 0 \\ \tau_x &= 0 & s_3 &= 0 \\ \tau_y &= 0 & s_4 &= 0 \end{aligned}$	$\begin{aligned} k_1 &= 0 & k_4 &= 0 \\ k_2 &= 0 & k_5 &= 0 \\ k_3 &= 0 & k_6 &= 0 \\ p_1 &= 0 & s_1 &= 0 \\ p_2 &= 0 & s_2 &= 0 \\ \tau_x &= 0 & s_3 &= 0 \\ \tau_y &= 0 & s_4 &= 0 \end{aligned}$	
Pattern pose	$M = \begin{pmatrix} 0.99619 & -7.60\mathrm{e}^{-3} & 0.086824 & -3 \\ 0 & 0.99619 & 0.087156 & -7 \\ -0.087156 & -0.086824 & 0.9924 & 0 \\ 0 & 0 & 0 & 1 \end{pmatrix}$	$M_{\text{cal.}} = \begin{pmatrix} 1 & 0 & 0 & 0 \\ 0 & 1 & 0 & -7.5 \\ 0 & 0 & 1 & 0 \\ 0 & 0 & 0 & 1 \end{pmatrix}$ Intrinsic calibration: Repr. error $= 0.036955\,\text{px}$, views $= 31$ Extrinsic calibration: Repr. error $= 25.039\,\text{px}$, views $= 31$	

Table A.14 – Test #13: Pattern misalignment 4, but without estimation of misalignment.

Ground truth	Calibrated camera model	Validation

Projection Parameters

$\text{width} = 1584,\ \text{height} = 2376$

$\text{fov}_y = 18.5°$

$c_x = 791.5,\ c_y = 1187.5$

$f_x \overset{!}{=} f_y,$

Estimate misalignment: False

$$P = \begin{pmatrix} 9.2103 & 0 & 0 & 0 \\ 0 & 6.1402 & 0 & 0 \\ 0 & 0 & -2.5 & -1050 \\ 0 & 0 & -1 & 0 \end{pmatrix}$$

$$K = \begin{pmatrix} 7295.1 & 0 & 791.49 \\ 0 & 7295.1 & 1187.6 \\ 0 & 0 & 1 \end{pmatrix}$$

$$P_{\text{cal.}} = \begin{pmatrix} 9.211 & 0 & 1.60\mathrm{e}^{-5} & 0 \\ 0 & 6.1407 & 5.07\mathrm{e}^{-5} & 0 \\ 0 & 0 & -1.2857 & -297.16 \\ 0 & 0 & -1 & 0 \end{pmatrix}$$

Camera pose

$$V = \begin{pmatrix} 1 & 0 & 0 & 0 \\ 0 & 0 & 1 & 0 \\ 0 & -1 & 0 & -520 \\ 0 & 0 & 0 & 1 \end{pmatrix}$$

$$(R|t) = \begin{pmatrix} 1 & 2.43\mathrm{e}^{-6} & -1.56\mathrm{e}^{-6} & 8.24\mathrm{e}^{-4} \\ -1.56\mathrm{e}^{-6} & -5.29\mathrm{e}^{-6} & -1 & -4.15\mathrm{e}^{-3} \\ -2.43\mathrm{e}^{-6} & 1 & -5.29\mathrm{e}^{-6} & 520.03 \end{pmatrix}$$

$$V_{\text{cal.}} = \begin{pmatrix} 1 & 2.43\mathrm{e}^{-6} & -1.56\mathrm{e}^{-6} & 8.24\mathrm{e}^{-4} \\ 1.56\mathrm{e}^{-6} & 5.29\mathrm{e}^{-6} & 1 & 4.15\mathrm{e}^{-3} \\ 2.43\mathrm{e}^{-6} & -1 & 5.29\mathrm{e}^{-6} & -520.03 \\ 0 & 0 & 0 & 1 \end{pmatrix}$$

Distortion

$k_1 = -4 \quad k_4 = 0$
$k_2 = 2 \quad k_5 = 0$
$k_3 = 1000 \quad k_6 = 0$
$p_1 = 0.3 \quad s_1 = 0$
$p_2 = -0.1 \quad s_2 = 0$
$\tau_x = 0 \quad s_3 = 0$
$\tau_y = 0 \quad s_4 = 0$

$k_1 = -4.0055 \quad k_4 = 0$
$k_2 = 2.5185 \quad k_5 = 0$
$k_3 = 985.59 \quad k_6 = 0$
$p_1 = 0.29999 \quad s_1 = 0$
$p_2 = -0.10001 \quad s_2 = 0$
$\tau_x = 0 \quad s_3 = 0$
$\tau_y = 0 \quad s_4 = 0$

Pattern pose

$$M = \begin{pmatrix} 1 & 0 & 0 & 0 \\ 0 & 1 & 0 & -7.5 \\ 0 & 0 & 1 & 0 \\ 0 & 0 & 0 & 1 \end{pmatrix}$$

$$M_{\text{cal.}} = \begin{pmatrix} 1 & 0 & 0 & 0 \\ 0 & 1 & 0 & -7.5 \\ 0 & 0 & 1 & 0 \\ 0 & 0 & 0 & 1 \end{pmatrix}$$

Intrinsic calibration:
Repr. error $= 0.047103\,\text{px}$, views $= 31$

Extrinsic calibration:
Repr. error $= 0.043816\,\text{px}$, views $= 31$

Table A.15 – Test #14: Distortion coefficients set 1: strong arbitrary distortion.

	Ground truth	Calibrated camera model	Validation
Projection Parameters	width = 1584, height = 2376 $\mathrm{fov}_y = 18.5°$ $c_x = 749.5,\ c_y = 1249.5$ $P = \begin{pmatrix} 9.2103 & 0 & 0.05303 & 0 \\ 0 & 6.1402 & 0.052189 & 0 \\ 0 & 0 & -2.5 & -1050 \\ 0 & 0 & -1 & 0 \end{pmatrix}$	$f_x \overset{!}{=} f_y,$ Estimate misalignment: False $K = \begin{pmatrix} 7294.4 & 0 & 749.42 \\ 0 & 7294.4 & 1249.9 \\ 0 & 0 & 1 \end{pmatrix}$	$P_{\mathrm{cal.}} = \begin{pmatrix} 9.2101 & 0 & 0.053131 & 0 \\ 0 & 6.1401 & 0.052518 & 0 \\ 0 & 0 & -1.2857 & -297.12 \\ 0 & 0 & -1 & 0 \end{pmatrix}$
Camera pose	$V = \begin{pmatrix} 1 & 0 & 0 & 0 \\ 0 & 0 & 1 & 0 \\ 0 & -1 & 0 & -520 \\ 0 & 0 & 0 & 1 \end{pmatrix}$	$(R\|t) = \begin{pmatrix} 1 & 1.10e^{-5} & -1.24e^{-7} & 5.69e^{-3} \\ -1.24e^{-7} & -4.02e^{-5} & -1 & -0.0278 \\ -1.10e^{-5} & 1 & -4.02e^{-5} & 519.96 \end{pmatrix}$	$V_{\mathrm{cal.}} = \begin{pmatrix} 1 & 1.10e^{-5} & -1.24e^{-7} & 5.69e^{-3} \\ 1.24e^{-7} & 4.02e^{-5} & 1 & 0.0278 \\ 1.10e^{-5} & -1 & 4.02e^{-5} & -519.96 \\ 0 & 0 & 0 & 1 \end{pmatrix}$
Distortion	$k_1 = 280.51 \quad k_4 = 279.08$ $k_2 = 1770.1 \quad k_5 = 1781.8$ $k_3 = 1053.5 \quad k_6 = -1008.6$ $p_1 = 0.06 \quad s_1 = 0.02$ $p_2 = 0 \quad s_2 = -0.24$ $\tau_x = 0 \quad s_3 = -0.05$ $\tau_y = 0 \quad s_4 = 0.34$	$k_1 = 284.24 \quad k_4 = 282.83$ $k_2 = 46.895 \quad k_5 = 60.384$ $k_3 = 864.73 \quad k_6 = -871.23$ $p_1 = 0.060015 \quad s_1 = 0.020004$ $p_2 = -1.15e^{-5} \quad s_2 = -0.24037$ $\tau_x = 0 \quad s_3 = -0.050062$ $\tau_y = 0 \quad s_4 = 0.34628$	
Pattern pose	$M = \begin{pmatrix} 1 & 0 & 0 & 0 \\ 0 & 1 & 0 & -7.5 \\ 0 & 0 & 1 & 0 \\ 0 & 0 & 0 & 1 \end{pmatrix}$	$M_{\mathrm{cal.}} = \begin{pmatrix} 1 & 0 & 0 & 0 \\ 0 & 1 & 0 & -7.5 \\ 0 & 0 & 1 & 0 \\ 0 & 0 & 0 & 1 \end{pmatrix}$ Intrinsic calibration: Repr. error = 0.044907 px, views = 31 Extrinsic calibration: Repr. error = 0.042569 px, views = 31	

Table A.16 – Test #15: Distortion coefficients set 2: moderate realistic distortion (values taken from table B.2, combination #29).

	Ground truth	Calibrated camera model	Validation
Projection Parameters	width $= 1584$, height $= 2376$ $\text{fov}_y = 18.5°$ $c_x = 791.5$, $c_y = 1187.5$ $P = \begin{pmatrix} 9.2103 & 0 & 0 & 0 \\ 0 & 6.1402 & 0 & 0 \\ 0 & 0 & -2.5 & -1050 \\ 0 & 0 & -1 & 0 \end{pmatrix}$	$f_x \overset{!}{=} f_y,$ Estimate misalignment: True $K = \begin{pmatrix} 7409 & 0 & 743.39 \\ 0 & 7409 & 1150 \\ 0 & 0 & 1 \end{pmatrix}$	$P_{\text{cal.}} = \begin{pmatrix} 9.3548 & 0 & 0.060743 & 0 \\ 0 & 6.2365 & -0.031593 & 0 \\ 0 & 0 & -1.2857 & -303.4 \\ 0 & 0 & -1 & 0 \end{pmatrix}$
Camera pose	$V = \begin{pmatrix} 1 & 0 & 0 & 7 \\ 0 & 0 & 1 & -4 \\ 0 & -1 & 0 & -520 \\ 0 & 0 & 0 & 1 \end{pmatrix}$	$(R\|t) = \begin{pmatrix} 0.99997 & 5.77\mathrm{e}^{-3} & 4.70\mathrm{e}^{-3} & 10.425 \\ 4.67\mathrm{e}^{-3} & 5.07\mathrm{e}^{-3} & -0.99998 & 5.9451 \\ -5.79\mathrm{e}^{-3} & 0.99997 & 5.05\mathrm{e}^{-3} & 530.95 \end{pmatrix}$	$V_{\text{cal.}} = \begin{pmatrix} 0.99997 & 5.77\mathrm{e}^{-3} & 4.70\mathrm{e}^{-3} & 10.425 \\ -4.67\mathrm{e}^{-3} & -5.07\mathrm{e}^{-3} & 0.99998 & -5.9451 \\ 5.79\mathrm{e}^{-3} & -0.99997 & -5.05\mathrm{e}^{-3} & -530.95 \\ 0 & 0 & 0 & 1 \end{pmatrix}$
Distortion	$k_1 = -4 \quad k_4 = 0$ $k_2 = 2 \quad k_5 = 0$ $k_3 = 1000 \quad k_6 = 0$ $p_1 = 0.3 \quad s_1 = 0$ $p_2 = -0.1 \quad s_2 = 0$ $\tau_x = 0 \quad s_3 = 0$ $\tau_y = 0 \quad s_4 = 0$	$k_1 = -4.3385 \quad k_4 = 0$ $k_2 = 3.3187 \quad k_5 = 0$ $k_3 = 915.77 \quad k_6 = 0$ $p_1 = 0.32926 \quad s_1 = 0$ $p_2 = -0.057165 \quad s_2 = 0$ $\tau_x = 0 \quad s_3 = 0$ $\tau_y = 0 \quad s_4 = 0$	
Pattern pose	$M = \begin{pmatrix} 0.99619 & -7.60\mathrm{e}^{-3} & 0.086824 & -3 \\ 0 & 0.99619 & 0.087156 & -7 \\ -0.087156 & -0.086824 & 0.9924 & 0 \\ 0 & 0 & 0 & 1 \end{pmatrix}$	$M_{\text{cal.}} = \begin{pmatrix} 0.99653 & -7.33\mathrm{e}^{-3} & 0.082925 & -3.0759 \\ 0 & 0.99612 & 0.088052 & -6.9829 \\ -0.083248 & -0.087747 & 0.99266 & 0 \\ 0 & 0 & 0 & 1 \end{pmatrix}$ Intrinsic calibration: Repr. error $= 3.5351$ px, views $= 31$ Extrinsic calibration: Repr. error $= 1.9373$ px, views $= 32$	

Table A.17 – Test #16: Almost everything combined with strong distortion (combination of tests #4, #12, #14).

	Ground truth	Calibrated camera model	Validation
Projection Parameters	width $= 1584$, height $= 2376$ $\text{fov}_y = 18.5°$ $c_x = 749.5,\ c_y = 1249.5$ $P = \begin{pmatrix} 9.2103 & 0 & 0.05303 & 0 \\ 0 & 6.1402 & 0.052189 & 0 \\ 0 & 0 & -2.5 & -1050 \\ 0 & 0 & -1 & 0 \end{pmatrix}$	$f_x \overset{!}{=} f_y,$ Estimate misalignment: True $K = \begin{pmatrix} 7294.3 & 0 & 749.57 \\ 0 & 7294.3 & 1249.4 \\ 0 & 0 & 1 \end{pmatrix}$	$P_{\text{cal.}} = \begin{pmatrix} 9.21 & 0 & 0.052947 & 0 \\ 0 & 6.14 & 0.052102 & 0 \\ 0 & 0 & -1.2857 & -297.14 \\ 0 & 0 & -1 & 0 \end{pmatrix}$
Camera pose	$V = \begin{pmatrix} 1 & 0 & 0 & 7 \\ 0 & 0 & 1 & -4 \\ 0 & -1 & 0 & -520 \\ 0 & 0 & 0 & 1 \end{pmatrix}$	$(R\|t) = \begin{pmatrix} 1 & 3.38\mathrm{e}^{-6} & -1.89\mathrm{e}^{-6} & 6.9952 \\ -1.89\mathrm{e}^{-6} & 1.82\mathrm{e}^{-5} & -1 & 3.3561 \\ -3.38\mathrm{e}^{-6} & 1 & 1.82\mathrm{e}^{-5} & 519.99 \end{pmatrix}$	$V_{\text{cal.}} = \begin{pmatrix} 1 & 3.38\mathrm{e}^{-6} & -1.89\mathrm{e}^{-6} & 6.9952 \\ 1.89\mathrm{e}^{-6} & -1.82\mathrm{e}^{-5} & 1 & -3.3561 \\ 3.38\mathrm{e}^{-6} & -1 & -1.82\mathrm{e}^{-5} & -519.99 \\ 0 & 0 & 0 & 1 \end{pmatrix}$
Distortion	$k_1 = 280.51 \quad k_4 = 279.08$ $k_2 = 1770.1 \quad k_5 = 1781.8$ $k_3 = 1053.5 \quad k_6 = -1008.6$ $p_1 = 0.06 \quad s_1 = 0.02$ $p_2 = 0 \quad s_2 = -0.24$ $\tau_x = 0 \quad s_3 = -0.05$ $\tau_y = 0 \quad s_4 = 0.34$	$k_1 = 303.19 \quad k_4 = 301.7$ $k_2 = 90.364 \quad k_5 = 102.43$ $k_3 = 896.62 \quad k_6 = -908.45$ $p_1 = 0.059988 \quad s_1 = 0.020074$ $p_2 = 1.52\mathrm{e}^{-5} \quad s_2 = -0.24693$ $\tau_x = 0 \quad s_3 = -0.049947$ $\tau_y = 0 \quad s_4 = 0.33617$	
Pattern pose	$M = \begin{pmatrix} 0.99619 & -7.60\mathrm{e}^{-3} & 0.086824 & -3 \\ 0 & 0.99619 & 0.087156 & -7 \\ -0.087156 & -0.086824 & 0.9924 & 0 \\ 0 & 0 & 0 & 1 \end{pmatrix}$	$M_{\text{cal.}} = \begin{pmatrix} 0.99619 & -7.60\mathrm{e}^{-3} & 0.086821 & -2.9429 \\ 0 & 0.99619 & 0.087159 & -6.9712 \\ -0.087153 & -0.086827 & 0.9924 & 0 \\ 0 & 0 & 0 & 1 \end{pmatrix}$ Intrinsic calibration: Repr. error $= 0.037699\,\text{px}$, views $= 31$ Extrinsic calibration: Repr. error $= 0.027585\,\text{px}$, views $= 31$	

Table A.18 – Test #17: Almost everything combined with realistic distortion (combination of tests #4, #12, #15).

	Ground truth	Calibrated camera model	Validation
Projection Parameters	width $= 1584$, height $= 2376$ fov$_y = 18.5°$ $c_x = 749.5$, $c_y = 1099.5$ $P = \begin{pmatrix} 9.2103 & 0 & 0.05303 & 0 \\ 0 & 6.1402 & -0.074074 & 0 \\ 0 & 0 & -2.5 & -1050 \\ 0 & 0 & -1 & 0 \end{pmatrix}$	$f_x \overset{!}{=} f_y,$ Estimate misalignment: True $K = \begin{pmatrix} 7434.1 & 0 & 693.55 \\ 0 & 7434.1 & 1065.5 \\ 0 & 0 & 1 \end{pmatrix}$	$P_{\text{cal.}} = \begin{pmatrix} 9.3865 & 0 & 0.12368 & 0 \\ 0 & 6.2577 & -0.10269 & 0 \\ 0 & 0 & -1.2857 & -304.12 \\ 0 & 0 & -1 & 0 \end{pmatrix}$
Camera pose	$V = \begin{pmatrix} 0.99624 & -0.07766 & -0.03836 & 7 \\ 0.039112 & 8.18e^{-3} & 0.9992 & -4 \\ -0.077284 & -0.99695 & 0.011187 & -520 \\ 0 & 0 & 0 & 1 \end{pmatrix}$	$(R|t) = \begin{pmatrix} 0.99694 & -0.070904 & -0.033002 & 10.944 \\ -0.033368 & -3.99e^{-3} & -0.99944 & 5.6849 \\ 0.070732 & 0.99748 & -6.34e^{-3} & 532.22 \end{pmatrix}$	$V_{\text{cal.}} = \begin{pmatrix} 0.99694 & -0.070904 & -0.033002 & 10.944 \\ 0.033368 & 3.99e^{-3} & 0.99944 & -5.6849 \\ -0.070732 & -0.99748 & 6.34e^{-3} & -532.22 \\ 0 & 0 & 0 & 1 \end{pmatrix}$
Distortion	$k_1 = -4 \quad k_4 = 0$ $k_2 = 2 \quad k_5 = 0$ $k_3 = 1000 \quad k_6 = 0$ $p_1 = 0.3 \quad s_1 = 0$ $p_2 = -0.1 \quad s_2 = 0$ $\tau_x = 0 \quad s_3 = 0$ $\tau_y = 0 \quad s_4 = 0$	$k_1 = -4.3602 \quad k_4 = 0$ $k_2 = 3.3285 \quad k_5 = 0$ $k_3 = 916.17 \quad k_6 = 0$ $p_1 = 0.3271 \quad s_1 = 0$ $p_2 = -0.052486 \quad s_2 = 0$ $\tau_x = 0 \quad s_3 = 0$ $\tau_y = 0 \quad s_4 = 0$	
Pattern pose	$M = \begin{pmatrix} 0.99619 & -7.60e^{-3} & 0.086824 & -3 \\ 0 & 0.99619 & 0.087156 & -7 \\ -0.087156 & -0.086824 & 0.9924 & 0 \\ 0 & 0 & 0 & 1 \end{pmatrix}$	$M_{\text{cal.}} = \begin{pmatrix} 0.99659 & -7.25e^{-3} & 0.082192 & -3.0775 \\ 0 & 0.99614 & 0.087826 & -6.9901 \\ -0.082511 & -0.087527 & 0.99274 & 0 \\ 0 & 0 & 0 & 1 \end{pmatrix}$ Intrinsic calibration: Repr. error $= 3.6376$ px, views $= 31$ Extrinsic calibration: Repr. error $= 2.0951$ px, views $= 31$	

Table A.19 – Test #18: Everything combined with strong distortion (combination of tests #3, #4, #8, #12, #14).

	Ground truth	Calibrated camera model	Validation
Parameters	width = 1584, height = 2376 $\mathrm{fov}_y = 18.5°$ $c_x = 749.5,\ c_y = 1099.5$	$f_x \overset{!}{=} f_y,$ Estimate misalignment: True	
Projection	$P = \begin{pmatrix} 9.2103 & 0 & 0.05303 & 0 \\ 0 & 6.1402 & -0.074074 & 0 \\ 0 & 0 & -2.5 & -1050 \\ 0 & 0 & -1 & 0 \end{pmatrix}$	$K = \begin{pmatrix} 7294.2 & 0 & 749.1 \\ 0 & 7294.2 & 1099.4 \\ 0 & 0 & 1 \end{pmatrix}$	$P_{\mathrm{cal.}} = \begin{pmatrix} 9.2098 & 0 & 0.053532 & 0 \\ 0 & 6.1399 & -0.074119 & 0 \\ 0 & 0 & -1.2857 & -297.14 \\ 0 & 0 & -1 & 0 \end{pmatrix}$
Camera pose	$V = \begin{pmatrix} 0.99624 & -0.07766 & -0.03836 & 7 \\ 0.039112 & 8.18\mathrm{e}^{-3} & 0.9992 & -4 \\ -0.077284 & -0.99695 & 0.011187 & -520 \\ 0 & 0 & 0 & 1 \end{pmatrix}$	$(R\vert t) = \begin{pmatrix} 0.99625 & -0.0776 & -0.038349 & 7.0038 \\ -0.039099 & -8.17\mathrm{e}^{-3} & -0.9992 & 3.3532 \\ 0.077225 & 0.99695 & -0.011172 & 520 \end{pmatrix}$	$V_{\mathrm{cal.}} = \begin{pmatrix} 0.99625 & -0.0776 & -0.038349 & 7.0038 \\ 0.039099 & 8.17\mathrm{e}^{-3} & 0.9992 & -3.3532 \\ -0.077225 & -0.99695 & 0.011172 & -520 \\ 0 & 0 & 0 & 1 \end{pmatrix}$
Distortion	$k_1 = 280.51 \quad k_4 = 279.08$ $k_2 = 1770.1 \quad k_5 = 1781.8$ $k_3 = 1053.5 \quad k_6 = -1008.6$ $p_1 = 0.06 \quad s_1 = 0.02$ $p_2 = 0 \quad s_2 = -0.24$ $\tau_x = 0 \quad s_3 = -0.05$ $\tau_y = 0 \quad s_4 = 0.34$	$k_1 = 329.25 \quad k_4 = 327.68$ $k_2 = 135.57 \quad k_5 = 141.61$ $k_3 = 839.56 \quad k_6 = -850.3$ $p_1 = 0.059974 \quad s_1 = 0.019906$ $p_2 = 8.60\mathrm{e}^{-7} \quad s_2 = -0.23277$ $\tau_x = 0 \quad s_3 = -0.050042$ $\tau_y = 0 \quad s_4 = 0.34769$	
Pattern pose	$M = \begin{pmatrix} 0.99619 & -7.60\mathrm{e}^{-3} & 0.086824 & -3 \\ 0 & 0.99619 & 0.087156 & -7 \\ -0.087156 & -0.086824 & 0.9924 & 0 \\ 0 & 0 & 0 & 1 \end{pmatrix}$	$M_{\mathrm{cal.}} = \begin{pmatrix} 0.9962 & -7.59\mathrm{e}^{-3} & 0.086812 & -2.9437 \\ 0 & 0.99619 & 0.087155 & -6.9714 \\ -0.087143 & -0.086824 & 0.99241 & 0 \\ 0 & 0 & 0 & 1 \end{pmatrix}$ Intrinsic calibration: Repr. error = 0.042499 px, views = 31 Extrinsic calibration: Repr. error = 0.030093 px, views = 31	

Table A.20 – Test #19: Everything combined with strong distortion (combination of tests #3, #4, #8, #12, #15).

Appendix B

Analysis of water distortions

The following tables tables B.1 to B.3 list the full results of the distortion analysis of three test cases with different positioning of the camera with respect to the bath walls (perpendicular or inclined) and with different media in the bath (air or water), see section 3.3.2. For each test case, the camera calibration routine was performed with 64 different combinations of OpenCV (Open Computer Vision) control flags, that determine which parameters of the camera model are varied or fixed, see eq. (3.3). A mark in columns 4 to 10 means that the respective flag was set. Resulting camera model parameters and reprojection errors are listed in the other columns. The row labeled 'no dist.' shows calibration results for the pinhole model without distortion. Results were sorted by values in column 3 (extrinsic reprojection error).

Combination #	$e_{extr.}$ [px]	$e_{extr.}$ [px]	fix $k_1 .. k_6 = 0$	fixAspectRatio	fixPrincipalPoint	rationalModel	thinPrismModel	tiltedModel	zeroTangentDist	f_x	f_y	c_x	c_y	k_1	k_2
29	0.3975	0.3746		x		x	x			5547.98	5547.98	767.75	1686.92	-81.47	3726.66
30	0.3997	0.3805		x		x		x		5578.28	5578.28	699.66	1971.83	1.04	22.30
20	0.4072	0.3868				x		x		5547.47	5513.50	1014.84	1388.54	61.20	747.82
51	0.4071	0.3870		x		x	x	x		5476.39	5476.39	660.82	971.24	76.51	566.13
6	0.4107	0.3876						x		5545.72	5511.77	1013.91	1396.13	-0.18	10.18
31	0.4088	0.3893		x			x	x		5479.13	5479.13	666.19	974.41	0.42	-16.92
56	0.4126	0.3905				x	x	x	x	5549.55	5516.05	1109.76	1432.85	18.54	709.54
5	0.4129	0.3911					x			5541.45	5505.08	959.40	1321.78	-0.18	12.52
19	0.4093	0.3912				x	x			5545.19	5508.23	969.47	1319.01	71.34	1450.55
10	0.4185	0.3914		x			x			5554.39	5554.39	778.54	1682.56	0.06	-0.58
40	0.4155	0.3936					x	x	x	5547.31	5513.76	1091.23	1428.79	-0.11	6.23
11	0.4186	0.3944		x				x		5578.00	5578.00	730.71	1942.47	0.23	-2.77
41	0.4285	0.4029				x	x	x		5517.02	5430.62	792.44	1185.19	0.04	0.98
21	0.4285	0.4029					x	x		5517.02	5430.63	792.44	1185.19	0.08	2.13
55	0.4286	0.4030			x	x	x	x		5517.29	5430.55	792.00	1188.00	0.04	1.23
37	0.4286	0.4030			x		x	x		5517.29	5430.56	792.00	1188.00	0.08	2.66
60	0.4311	0.4048		x	x	x	x	x		5485.71	5485.71	792.00	1188.00	-34.28	27.38
47	0.4312	0.4049		x	x		x	x		5485.64	5485.64	792.00	1188.00	0.08	0.64
4	0.4512	0.4272				x				5530.36	5498.35	822.88	1279.61	0.36	21.81
0	0.4512	0.4272								5530.34	5498.33	822.86	1279.56	-0.17	15.11
35	0.4542	0.4548			x	x	x			5520.22	5485.35	792.00	1188.00	-34.76	-103.17
14	0.4544	0.4552			x		x			5520.14	5485.34	792.00	1188.00	0.06	1.62
62	0.4552	0.4573			x	x	x	x	x	5519.98	5485.21	792.00	1188.00	-34.34	-119.23
54	0.4554	0.4577			x		x	x	x	5519.89	5485.20	792.00	1188.00	0.06	1.59
39	0.4460	0.4615				x		x	x	5505.02	5469.56	671.17	957.81	193.49	381.91
17	0.4556	0.4626					x		x	5515.82	5481.30	762.00	1165.01	0.17	-5.07
50	0.4386	0.4626		x			x	x	x	5609.14	5609.14	428.98	2170.86	0.09	-0.58
38	0.4556	0.4626				x	x		x	5515.80	5481.28	762.00	1165.01	4.61	-3.08
18	0.4478	0.4633						x	x	5510.36	5474.93	670.87	962.59	0.40	-14.88
15	0.4682	0.4655			x			x		5519.51	5484.81	792.00	1188.00	0.09	-2.40
36	0.4682	0.4655			x	x		x		5519.54	5484.83	792.00	1188.00	-3.92	1.13
2	0.4705	0.4680			x					5520.23	5485.35	792.00	1188.00	0.09	-2.25
13	0.4705	0.4680			x	x				5520.24	5485.37	792.00	1188.00	-3.34	1.35
52	0.4838	0.4814			x	x	x		x	5517.44	5483.46	792.00	1188.00	-32.11	-202.54
33	0.4840	0.4823			x		x		x	5517.45	5483.55	792.00	1188.00	0.07	1.35
16	0.4895	0.4909				x			x	5514.51	5479.45	755.32	1148.86	-13.05	160.03
3	0.4895	0.4909							x	5514.42	5479.35	755.28	1148.85	0.23	-9.86
34	0.4977	0.4952			x			x	x	5520.16	5485.14	792.00	1188.00	0.10	-2.32
53	0.4977	0.4952			x	x		x	x	5520.19	5485.17	792.00	1188.00	-7.75	6.24
12	0.5792	0.5810			x				x	5512.12	5479.37	792.00	1188.00	0.14	-5.43
32	0.5791	0.5812			x	x			x	5512.08	5479.36	792.00	1188.00	-27.96	5.72
28	0.5166	0.6273		x				x	x	5613.04	5613.04	-381.90	1305.95	-0.04	1.19
61	0.4267	0.6460		x		x	x	x	x	5598.92	5598.92	399.63	2154.75	0.11	11.11
49	0.5351	0.6756		x		x		x	x	5580.73	5580.73	-256.76	1280.83	-2.49	-56.01
9	0.5848	0.6919		x		x				5503.94	5503.94	803.98	1482.03	-74.03	2027.12
1	0.5873	0.7097		x						5505.15	5505.15	805.55	1475.65	-0.06	4.12
48	0.7208	1.3935		x		x	x		x	5439.38	5439.38	757.79	1187.16	-30.82	-200.51
27	0.7212	1.4047		x			x		x	5437.68	5437.68	757.99	1186.31	0.22	-12.82
59	0.7224	1.4073		x	x		x	x	x	5438.94	5438.94	792.00	1188.00	0.20	-12.35
63	0.7218	1.4089		x	x	x	x	x	x	5439.83	5439.83	792.00	1188.00	-16.41	-718.92
24	0.7239	1.4123		x	x		x			5438.80	5438.80	792.00	1188.00	0.20	-12.32
45	0.7233	1.4139		x	x	x	x			5439.69	5439.69	792.00	1188.00	-16.68	-707.89
43	0.7357	1.4175		x	x		x		x	5438.92	5438.92	792.00	1188.00	0.19	-11.87
25	0.7347	1.4186		x	x			x		5438.24	5438.24	792.00	1188.00	0.23	-15.65
7	0.7387	1.4188		x	x					5438.69	5438.69	792.00	1188.00	0.22	-15.49
57	0.7353	1.4190		x	x	x	x		x	5439.74	5439.74	792.00	1188.00	-15.35	-754.03
46	0.7340	1.4207		x	x	x		x		5439.43	5439.43	792.00	1188.00	-19.07	-564.27
23	0.7381	1.4208		x	x	x				5439.86	5439.86	792.00	1188.00	-18.66	-573.20
44	0.7552	1.4315		x	x			x	x	5438.36	5438.36	792.00	1188.00	0.23	-15.95
58	0.7545	1.4332		x	x	x		x	x	5439.43	5439.43	792.00	1188.00	-16.62	-638.11
8	0.7519	1.4351		x					x	5433.93	5433.93	753.94	1162.86	0.31	-19.57
no dist.	0.8023	1.4592	x	x	x				x	5444.73	5444.73	792.00	1188.00	0.00	0.00
22	0.7885	1.4715		x	x				x	5436.79	5436.79	792.00	1188.00	0.25	-17.35
42	0.7875	1.4731		x	x	x			x	5437.68	5437.68	792.00	1188.00	-13.03	-796.64
26	0.7511	1.5010		x		x			x	5436.16	5436.16	753.68	1162.99	-27.82	-364.97

Table B.1 – Full results of distortion test case 1: camera perpendicular, medium air. Rows were sorted by values in column 3 (extrinsic reprojection error $e_{\mathrm{extr.}}$).

Combination #	p_1	p_2	k_3	k_4	k_5	k_6	s_1	s_2	s_3	s_4	τ_x	τ_y
29	0.05	0.00	539.75	-81.70	3743.13	40.28	0.01	-0.11	-0.05	0.15		
30	0.02	0.00	1090.48	0.73	27.09	1038.72	0.00	0.00	0.00	0.00	0.14	0.02
20	0.01	0.01	-52.41	61.65	729.25	52.21	0.00	0.00	0.00	0.00	0.03	-0.04
51	-0.02	-0.06	170.46	75.78	580.60	-161.69	0.06	0.07	0.02	-0.11	-0.01	-0.11
6	0.01	0.01	-117.33	0.00	0.00	0.00	0.00	0.00	0.00	0.00	0.03	-0.04
31	-0.02	-0.06	243.49	0.00	0.00	0.00	0.06	0.07	0.02	-0.09	-0.01	-0.11
56	0.00	0.00	113.90	18.74	700.92	172.59	0.01	-0.02	0.01	-0.15	0.05	-0.07
5	0.02	0.02	-183.38	0.00	0.00	0.00	-0.01	-0.09	0.00	-0.22		
19	0.02	0.02	-137.21	71.85	1424.40	148.72	-0.01	-0.10	0.00	-0.26		
10	0.05	0.00	18.30	0.00	0.00	0.00	0.01	-0.10	-0.05	0.14		
40	0.00	0.00	-67.29	0.00	0.00	0.00	0.01	-0.02	0.01	-0.14	0.05	-0.07
11	0.02	0.00	18.93	0.00	0.00	0.00	0.00	0.00	0.00	0.00	0.14	0.01
41	0.07	0.00	-47.93	-0.04	-1.14	33.06	0.01	-0.08	-0.05	-0.62	-0.14	-0.01
21	0.07	0.00	-81.13	0.00	0.00	0.00	0.01	-0.08	-0.05	-0.62	-0.14	-0.01
55	0.07	0.00	-53.71	-0.04	-1.42	38.55	0.01	-0.09	-0.05	-0.62	-0.14	-0.01
37	0.07	0.00	-92.33	0.00	0.00	0.00	0.01	-0.09	-0.05	-0.62	-0.14	-0.01
60	0.01	-0.05	0.19	-34.35	30.03	1.03	0.06	-0.09	0.01	-0.62	-0.01	-0.11
47	0.01	-0.05	-27.86	0.00	0.00	0.00	0.06	-0.08	0.01	-0.62	-0.01	-0.11
4	0.01	0.01	-119.00	0.53	6.63	120.00	0.00	0.00	0.00	0.00	0.00	0.00
0	0.01	0.01	-240.13									
35	0.00	0.00	-2.01	-34.84	-100.26	1.12	0.00	-0.04	0.02	-0.62		
14	0.00	0.00	-54.52	0.00	0.00	0.00	0.00	-0.03	0.02	-0.61		
62	0.00	0.00	-2.18	-34.42	-116.32	1.26	0.00	-0.04	0.02	-0.62	0.00	-0.01
54	0.00	0.00	-54.31	0.00	0.00	0.00	0.00	-0.03	0.02	-0.62	0.00	-0.01
39	0.00	0.00	-11.71	192.42	377.85	13.93	0.00	0.00	0.00	0.00	-0.04	0.02
17	0.00	0.00	77.85	0.00	0.00	0.00	0.00	0.01	0.02	-0.65		
50	0.00	0.00	3.57	0.00	0.00	0.00	0.01	-0.06	0.02	0.04	0.20	0.07
38	0.00	0.00	36.20	4.45	1.56	-37.12	0.00	0.01	0.02	-0.65		
18	0.00	0.00	193.61	0.00	0.00	0.00	0.00	0.00	0.00	0.00	-0.04	0.02
15	0.01	0.00	95.81	0.00	0.00	0.00	0.00	0.00	0.00	0.00	-0.01	0.00
36	0.01	0.00	42.51	-4.01	3.55	-42.60	0.00	0.00	0.00	0.00	-0.01	0.00
2	0.00	0.00	90.74									
13	0.00	0.00	41.02	-3.43	3.63	-41.05	0.00	0.00	0.00	0.00	0.00	0.00
52	0.00	0.00	7.19	-32.20	-198.74	-8.75	0.01	-0.04	0.02	-0.67		
33	0.00	0.00	-64.32	0.00	0.00	0.00	0.01	-0.03	0.02	-0.67		
16	0.00	0.00	101.44	-13.26	170.65	-102.10	0.00	0.00	0.00	0.00	0.00	0.00
3	0.00	0.00	190.18									
34	0.00	0.00	81.26	0.00	0.00	0.00	0.00	0.00	0.00	0.00	0.01	-0.01
53	0.00	0.00	33.05	-7.84	8.73	-33.41	0.00	0.00	0.00	0.00	0.01	-0.01
12	0.00	0.00	100.44									
32	0.00	0.00	54.25	-28.09	12.68	-54.82	0.00	0.00	0.00	0.00	0.00	0.00
28	0.00	0.00	-4.60	0.00	0.00	0.00	0.00	0.00	0.00	0.00	0.03	0.23
61	0.00	0.00	912.89	-0.07	13.90	886.70	0.01	-0.06	0.02	0.03	0.19	0.07
49	0.00	0.00	-41.66	-2.49	-57.01	-33.21	0.00	0.00	0.00	0.00	0.03	0.21
9	0.03	0.00	87.09	-74.07	2032.01	-60.28	0.00	0.00	0.00	0.00	0.00	0.00
1	0.03	0.00	-26.47									
48	0.00	0.00	28.22	-30.93	-194.83	-36.12	0.00	0.07	0.02	-0.60		
27	0.00	0.00	291.19	0.00	0.00	0.00	0.00	0.08	0.02	-0.59		
59	0.00	0.00	309.39	0.00	0.00	0.00	0.00	-0.05	0.02	-0.55	0.00	-0.01
63	0.00	0.00	19.57	-16.52	-713.30	-37.47	0.00	-0.03	0.02	-0.55	0.00	-0.01
24	0.00	0.00	308.73	0.00	0.00	0.00	0.00	-0.05	0.02	-0.55		
45	0.00	0.00	19.52	-16.79	-702.29	-37.32	0.00	-0.03	0.02	-0.54		
43	0.00	0.00	293.21	0.00	0.00	0.00	0.01	-0.05	0.02	-0.56		
25	0.01	0.00	431.32	0.00	0.00	0.00	0.00	0.00	0.00	0.00	-0.01	0.00
7	0.00	0.00	425.75									
57	0.00	0.00	21.40	-15.46	-748.25	-39.26	0.00	-0.03	0.02	-0.56		
46	0.00	0.00	18.73	-19.18	-558.95	-44.16	0.00	0.00	0.00	0.00	-0.01	0.00
23	0.00	0.00	20.45	-18.77	-567.74	-45.70	0.00	0.00	0.00	0.00	0.00	0.00
44	0.00	0.00	423.90	0.00	0.00	0.00	0.00	0.00	0.00	0.00	0.01	-0.01
58	0.00	0.00	34.55	-16.74	-631.32	-60.36	0.00	0.00	0.00	0.00	0.01	-0.01
8	0.00	0.00	458.11									
no dist.	0.00	0.00	0.00									
22	0.00	0.00	424.08									
42	0.00	0.00	66.90	-13.19	-786.91	-87.26	0.00	0.00	0.00	0.00	0.00	0.00
26	0.00	0.00	48.56	-27.98	-356.09	-65.84	0.00	0.00	0.00	0.00	0.00	0.00

Table B.1 – (continued).

Combination #	$e_{intr.}$ [px]	$e_{extr.}$ [px]	fix $k_1..k_6=0$	fixAspectRatio	fixPrincipalPoint	rationalModel	thinPrismModel	tiltedModel	zeroTangentDist	f_x	f_y	c_x	c_y	k_1	k_2
30	0.4201	0.4000		x		x		x		7459.80	7459.80	641.23	2065.65	2.41	83.20
29	0.4321	0.4085		x		x	x			7425.35	7425.35	750.65	1735.52	280.51	1770.08
51	0.4407	0.4118		x		x	x	x		7380.72	7380.72	1028.26	1374.40	121.95	342.17
31	0.4439	0.4141		x			x	x		7379.83	7379.83	1030.38	1391.57	-0.17	24.20
10	0.4488	0.4175		x			x			7437.30	7437.30	760.67	1728.27	0.19	-2.41
20	0.4433	0.4181				x		x		7424.88	7378.38	1006.20	1377.25	95.85	205.82
6	0.4458	0.4185						x		7422.27	7375.90	1001.64	1379.71	-0.23	29.93
11	0.4502	0.4191		x				x		7461.87	7461.87	692.60	2013.98	0.48	-7.80
40	0.4520	0.4212					x	x	x	7448.06	7400.89	1201.36	1638.69	0.01	4.74
5	0.4470	0.4222					x			7417.49	7371.12	956.06	1330.57	-0.23	33.34
19	0.4445	0.4228				x	x			7421.57	7374.42	966.02	1326.37	127.21	109.73
21	0.4495	0.4251					x	x		7353.43	7323.71	722.03	1049.39	0.74	-50.96
41	0.4495	0.4251				x	x	x		7353.04	7324.03	721.67	1049.49	-4.57	-120.53
55	0.4663	0.4354			x	x	x	x		7382.15	7343.93	792.00	1188.00	0.07	3.40
37	0.4663	0.4354			x		x	x		7382.14	7343.94	792.00	1188.00	0.12	10.79
60	0.4670	0.4361		x	x	x	x	x		7347.30	7347.30	792.00	1188.00	220.35	47.17
47	0.4671	0.4363		x	x		x	x		7346.37	7346.37	792.00	1188.00	0.12	10.14
56	0.4532	0.4378				x	x	x	x	7419.51	7373.70	1059.89	1399.70	75.59	1506.94
4	0.4782	0.4427				x				7409.61	7367.33	854.18	1292.18	31.81	75.62
0	0.4784	0.4440								7408.42	7366.08	853.12	1290.71	-0.26	42.75
35	0.4830	0.4787			x	x	x			7393.41	7347.03	792.00	1188.00	204.61	42.69
14	0.4831	0.4788			x		x			7392.55	7346.15	792.00	1188.00	0.11	11.69
62	0.4859	0.4872			x	x	x	x	x	7393.02	7346.20	792.00	1188.00	198.79	41.27
54	0.4860	0.4874			x		x	x	x	7392.19	7345.34	792.00	1188.00	0.11	11.65
15	0.4978	0.4891			x			x		7391.88	7345.71	792.00	1188.00	0.16	1.84
36	0.4978	0.4892			x	x		x		7391.91	7345.74	792.00	1188.00	-29.73	3.20
2	0.4986	0.4904			x					7391.54	7345.51	792.00	1188.00	0.16	1.68
13	0.4986	0.4905			x	x				7391.57	7345.54	792.00	1188.00	-29.75	3.26
38	0.4809	0.4907				x	x		x	7378.81	7332.65	752.42	1111.77	-29.87	-4.94
17	0.4811	0.4909					x		x	7378.36	7332.23	752.10	1111.82	0.54	-34.86
39	0.4713	0.4912				x		x	x	7364.54	7317.03	675.43	954.77	239.22	85.45
18	0.4727	0.4912						x	x	7370.95	7323.47	678.02	960.94	0.76	-44.64
16	0.5065	0.5141				x			x	7377.83	7330.64	741.09	1095.62	-16.61	-9.01
3	0.5066	0.5141							x	7377.62	7330.43	741.01	1095.68	0.57	-32.63
34	0.5302	0.5307			x			x	x	7392.75	7345.17	792.00	1188.00	0.16	1.54
53	0.5302	0.5307			x	x		x	x	7392.72	7345.15	792.00	1188.00	-28.33	-3.02
50	0.5051	0.5544		x				x	x	7481.28	7481.28	654.28	2228.45	0.07	-0.75
61	0.4990	0.5545		x		x		x	x	7475.69	7475.69	606.94	2227.03	-0.33	-46.26
52	0.5639	0.5550			x	x			x	7382.99	7339.82	792.00	1188.00	30.50	46.06
33	0.5639	0.5550			x				x	7382.98	7339.79	792.00	1188.00	0.15	7.05
9	0.5726	0.6178		x		x				7386.29	7386.29	809.73	1551.60	-106.68	4169.29
1	0.5773	0.6387		x						7386.89	7386.89	812.14	1543.02	0.02	6.15
49	0.5730	0.7087		x		x		x	x	7432.30	7432.30	-275.68	1257.41	71.20	-149.49
28	0.5727	0.7162		x				x	x	7418.17	7418.17	-258.92	1253.02	-0.10	5.27
12	0.7249	0.7385			x				x	7372.81	7332.18	792.00	1188.00	0.32	-17.98
32	0.7245	0.7395			x	x			x	7372.30	7331.81	792.00	1188.00	-51.09	-197.62
25	0.7401	1.4735		x	x				x	7283.18	7283.18	792.00	1188.00	0.32	-24.63
46	0.7401	1.4735		x	x	x			x	7283.18	7283.18	792.00	1188.00	0.15	-14.43
24	0.7306	1.4747		x	x		x			7283.42	7283.42	792.00	1188.00	0.28	-16.31
7	0.7413	1.4753		x	x					7283.19	7283.19	792.00	1188.00	0.32	-24.69
45	0.7302	1.4757		x	x	x	x			7284.19	7284.19	792.00	1188.00	-54.39	-142.21
59	0.7322	1.4758		x	x			x	x	7283.01	7283.01	792.00	1188.00	0.28	-17.00
63	0.7317	1.4767		x	x	x		x	x	7283.78	7283.78	792.00	1188.00	-49.77	-415.18
23	0.7408	1.4768		x	x	x				7284.56	7284.56	792.00	1188.00	-53.34	-86.12
43	0.7605	1.4866		x	x		x		x	7282.56	7282.56	792.00	1188.00	0.30	-18.11
57	0.7599	1.4879		x	x	x	x		x	7283.12	7283.12	792.00	1188.00	-46.78	-616.58
48	0.7269	1.4938		x		x		x	x	7274.36	7274.36	745.01	1129.44	4.92	-16.32
27	0.7269	1.4938		x				x	x	7274.42	7274.42	745.04	1129.46	0.51	-33.88
44	0.7710	1.5126		x	x			x	x	7281.48	7281.48	792.00	1188.00	0.34	-27.33
58	0.7704	1.5137		x	x	x		x	x	7282.74	7282.74	792.00	1188.00	-46.11	-451.98
26	0.7561	1.5139		x		x			x	7270.47	7270.47	736.59	1113.39	37.07	129.35
8	0.7561	1.5141		x					x	7270.91	7270.91	736.65	1113.40	0.52	-30.00
no dist.	0.9011	1.5695	x	x	x				x	7299.05	7299.05	792.00	1188.00	0.00	0.00
22	0.8675	1.5896		x	x				x	7279.32	7279.32	792.00	1188.00	0.45	-39.57
42	0.8658	1.6660		x	x	x			x	7280.38	7280.38	792.00	1188.00	-47.95	-468.42

Table B.2 – **Full results of distortion test case 2: camera perpendicular, medium water.** Rows were sorted by values in column 3 (extrinsic reprojection error $e_{extr.}$).

Combination #	p_1	p_2	k_3	k_4	k_5	k_6	s_1	s_2	s_3	s_4	τ_x	τ_y
30	0.03	0.00	6535.71	1.70	100.29	6191.67	0.00	0.00	0.00	0.00	0.14	0.03
29	0.06	0.00	1053.54	279.08	1781.76	-1008.64	0.02	-0.24	-0.05	0.34		
51	0.02	-0.03	-436.39	122.63	275.40	438.89	0.04	-0.15	0.00	-0.30	0.01	-0.11
31	0.02	-0.03	-472.88	0.00	0.00	0.00	0.04	-0.14	-0.01	-0.23	0.01	-0.11
10	0.06	0.00	89.10	0.00	0.00	0.00	0.01	-0.23	-0.05	0.35		
20	0.01	0.01	-348.71	96.49	144.69	352.53	0.00	0.00	0.00	0.00	0.03	-0.03
6	0.01	0.01	-577.73	0.00	0.00	0.00	0.00	0.00	0.00	0.00	0.03	-0.03
11	0.03	0.00	80.32	0.00	0.00	0.00	0.00	0.00	0.00	0.00	0.13	0.02
40	0.00	0.00	-54.76	0.00	0.00	0.00	0.02	-0.15	0.02	-0.23	0.09	-0.08
5	0.02	0.02	-776.61	0.00	0.00	0.00	-0.01	-0.25	-0.01	-0.40		
19	0.02	0.02	-688.99	127.94	29.00	692.29	-0.01	-0.26	0.00	-0.47		
21	-0.02	0.03	1490.16	0.00	0.00	0.00	-0.03	0.05	0.03	-0.94	0.02	0.07
41	-0.02	0.03	709.67	-5.29	-69.71	-717.07	-0.03	0.05	0.03	-0.94	0.02	0.07
55	-0.01	0.03	-211.56	-0.04	-7.36	211.71	-0.02	-0.27	0.03	-1.24	0.02	0.05
37	-0.01	0.03	-423.58	0.00	0.00	0.00	-0.02	-0.27	0.03	-1.24	0.02	0.05
60	0.01	-0.05	-433.87	220.36	-17.58	436.00	0.06	-0.28	0.01	-1.22	-0.01	-0.11
47	0.01	-0.05	-395.28	0.00	0.00	0.00	0.06	-0.28	0.01	-1.23	-0.01	-0.11
56	0.00	0.00	-232.91	76.08	1461.37	235.12	0.01	-0.10	0.02	-0.33	0.05	-0.06
4	0.02	0.01	-540.97	32.18	21.55	546.56	0.00	0.00	0.00	0.00	0.00	0.00
0	0.02	0.01	-1050.36									
35	0.01	0.00	-483.45	204.64	-20.73	485.35	0.00	-0.20	0.01	-1.22		
14	0.01	0.00	-462.01	0.00	0.00	0.00	0.00	-0.20	0.01	-1.23		
62	0.00	0.00	-481.22	198.81	-20.58	483.11	0.01	-0.19	0.02	-1.23	0.01	-0.01
54	0.00	0.00	-462.20	0.00	0.00	0.00	0.01	-0.19	0.02	-1.24	0.01	-0.01
15	0.01	0.01	144.02	0.00	0.00	0.00	0.00	0.00	0.00	0.00	0.00	0.00
36	0.01	0.01	-25.47	-29.88	4.50	25.89	0.00	0.00	0.00	0.00	0.00	0.00
2	0.01	0.00	152.82									
13	0.01	0.00	-23.37	-29.90	4.62	23.80	0.00	0.00	0.00	0.00	0.00	0.00
38	0.00	0.00	522.91	-30.36	33.78	-527.30	0.00	-0.08	0.02	-1.33		
17	0.00	0.00	1116.67	0.00	0.00	0.00	0.00	-0.07	0.02	-1.32		
39	0.00	0.00	219.08	237.58	86.28	-220.62	0.00	0.00	0.00	0.00	-0.03	0.01
18	0.00	0.00	990.40	0.00	0.00	0.00	0.00	0.00	0.00	0.00	-0.03	0.01
16	0.00	0.00	434.86	-17.15	28.07	-437.11	0.00	0.00	0.00	0.00	0.00	0.00
3	0.00	0.00	767.56									
34	0.00	0.00	108.17	0.00	0.00	0.00	0.00	0.00	0.00	0.00	0.02	-0.01
53	0.00	0.00	-13.55	-28.49	-0.79	13.77	0.00	0.00	0.00	0.00	0.02	-0.01
50	0.00	0.00	16.77	0.00	0.00	0.00	0.01	-0.10	0.02	0.11	0.17	0.01
61	0.00	0.00	1690.67	-0.46	-43.42	1647.05	0.01	-0.11	0.02	0.11	0.17	0.02
52	0.00	0.00	-302.47	30.36	30.85	303.01	0.01	-0.21	0.03	-1.50		
33	0.00	0.00	-436.68	0.00	0.00	0.00	0.01	-0.21	0.04	-1.51		
9	0.03	0.01	386.35	-106.82	4187.60	-362.95	0.00	0.00	0.00	0.00	0.00	0.00
1	0.03	0.01	-22.92									
49	0.00	0.00	574.55	71.61	-161.76	598.33	0.00	0.00	0.00	0.00	0.03	0.17
28	0.00	0.00	-39.92	0.00	0.00	0.00	0.00	0.00	0.00	0.00	0.03	0.17
12	0.00	0.00	462.19									
32	0.00	0.00	309.19	-51.39	-170.27	-310.91	0.00	0.00	0.00	0.00	0.00	0.00
25	0.01	0.01	1198.38	0.00	0.00	0.00	0.00	0.00	0.00	0.00	0.00	0.00
46	0.01	0.01	527.07	-0.17	10.14	-666.56	0.00	0.00	0.00	0.00	0.00	0.00
24	0.00	0.00	691.04	0.00	0.00	0.00	0.00	-0.23	0.02	-1.12		
7	0.01	0.00	1199.56									
45	0.00	0.00	55.45	-54.58	-129.22	-56.53	0.00	-0.20	0.02	-1.12		
59	0.00	0.00	706.70	0.00	0.00	0.00	0.01	-0.22	0.02	-1.15	0.01	-0.01
63	0.00	0.00	45.18	-49.97	-402.20	-48.34	0.01	-0.20	0.02	-1.15	0.01	-0.01
23	0.01	0.00	32.80	-53.52	-75.50	-31.41	0.00	0.00	0.00	0.00	0.00	0.00
43	0.00	0.00	658.36	0.00	0.00	0.00	0.01	-0.23	0.03	-1.34		
57	0.00	0.00	101.00	-47.01	-600.12	-104.23	0.01	-0.20	0.03	-1.33		
48	0.00	0.00	514.82	4.41	16.95	-517.60	0.00	0.16	0.02	-1.61		
27	0.00	0.00	1030.89	0.00	0.00	0.00	0.00	0.16	0.02	-1.61		
44	0.00	0.00	1230.36	0.00	0.00	0.00	0.00	0.00	0.00	0.00	0.01	-0.01
58	0.00	0.00	84.01	-46.32	-437.96	-87.46	0.00	0.00	0.00	0.00	0.01	-0.01
26	0.00	0.00	256.06	36.50	150.54	-255.07	0.00	0.00	0.00	0.00	0.00	0.00
8	0.00	0.00	697.19									
no dist.	0.00	0.00	0.00									
22	0.00	0.00	1420.54									
42	0.00	0.00	342.06	-48.25	-439.49	-344.10	0.00	0.00	0.00	0.00	0.00	0.00

Table B.2 – (continued).

Combination #	$e_{intr.}$ [px]	$e_{extr.}$ [px]	fix $k_1 .. k_6 = 0$	fixAspectRatio	fixPrincipalPoint	rationalModel	thinPrismModel	tiltedModel	zeroTangentDist	f_x	f_y	c_x	c_y	k_1	k_2
29	0.5259	0.4256	x			x	x			7643.46	7643.46	765.68	1832.94	-129.01	5315.94
10	0.5434	0.4314	x				x			7640.82	7640.82	772.88	1820.01	0.10	4.38
6	0.5491	0.4341						x		7485.40	7286.16	631.08	1054.50	0.65	-39.49
51	0.5436	0.4361	x			x	x	x		7363.71	7363.71	593.18	1102.58	211.22	164.90
20	0.5476	0.4384				x		x		7478.85	7279.40	625.65	1062.03	291.06	72.71
56	0.5425	0.4391				x	x	x	x	7489.28	7314.15	542.08	1102.87	399.27	369.89
31	0.5461	0.4399	x				x	x		7370.94	7370.94	620.21	1110.09	0.69	-47.46
40	0.5456	0.4409				x	x	x	x	7497.36	7322.76	541.56	1089.37	0.58	-32.62
41	0.5530	0.4431			x	x	x			7499.17	7359.60	769.54	1174.99	0.12	-36.29
21	0.5530	0.4432				x	x			7499.14	7360.16	769.68	1175.13	0.60	-47.59
55	0.5553	0.4453		x	x	x	x			7505.65	7320.08	792.00	1188.00	0.18	-26.40
37	0.5553	0.4453		x		x	x			7505.65	7320.50	792.00	1188.00	0.53	-40.95
19	0.5597	0.4478				x	x			7524.16	7413.49	871.68	1243.86	-54.36	-535.47
60	0.5560	0.4481	x	x	x	x	x			7394.46	7394.46	792.00	1188.00	-49.80	-182.92
47	0.5566	0.4491	x	x		x	x			7392.73	7392.73	792.00	1188.00	0.56	-42.62
5	0.5487	0.4585				x				7578.52	7506.22	854.83	1491.92	-0.26	42.20
15	0.5659	0.4696		x			x			7501.57	7293.20	792.00	1188.00	0.50	-32.60
36	0.5654	0.4705		x	x		x			7503.33	7294.16	792.00	1188.00	-59.04	-11.09
50	0.5584	0.4942	x			x	x	x		7493.21	7493.21	-127.55	-185.91	0.31	-0.50
18	0.5895	0.5053				x	x	x		7501.40	7321.82	578.36	1257.65	0.50	-32.53
39	0.5898	0.5211			x	x	x	x		7481.08	7307.22	596.58	1167.30	178.86	187.16
35	0.5619	0.5310		x	x	x				7505.69	7390.88	792.00	1188.00	-51.46	-160.19
14	0.5625	0.5316		x		x				7503.54	7389.27	792.00	1188.00	0.54	-39.79
54	0.5684	0.5850		x		x	x	x		7504.26	7321.70	792.00	1188.00	0.53	-40.54
53	0.6158	0.6294		x	x		x	x		7487.04	7311.07	792.00	1188.00	-57.40	49.64
34	0.6162	0.6305		x			x	x		7485.85	7310.43	792.00	1188.00	0.50	-31.25
3	0.6431	0.6310						x		7458.41	7290.05	758.70	27.10	-0.24	3.18
16	0.6435	0.6322		x				x		7456.92	7288.86	758.25	46.15	-0.40	46.66
38	0.5598	0.6429		x	x			x		7473.17	7338.67	771.29	356.36	-116.80	4758.21
17	0.5648	0.6460		x				x		7474.87	7343.80	772.33	437.24	0.01	7.93
62	0.5678	0.7709		x	x	x	x	x		7506.32	7323.10	792.00	1188.00	-49.60	-159.37
61	0.6142	0.7880	x		x	x	x	x		7413.12	7413.12	116.97	175.36	-0.32	-151.50
4	0.6794	0.8209			x					7404.46	7276.87	779.09	547.91	0.20	8.23
0	0.6794	0.8209								7404.44	7276.87	779.05	547.95	0.39	0.74
1	0.6615	1.1381	x							7424.98	7424.98	722.98	190.44	0.34	4.36
9	0.6881	1.2826	x		x					7419.90	7419.90	730.49	213.80	-0.52	-204.99
13	1.0698	1.4119		x	x					7357.23	7277.83	792.00	1188.00	1.15	31.67
2	1.0698	1.4121		x						7357.24	7277.85	792.00	1188.00	0.45	-14.09
48	0.8228	1.8376	x		x	x			x	7421.55	7421.55	776.46	109.05	2.14	43.56
27	0.8253	1.8496	x			x			x	7422.32	7422.32	775.98	97.95	-0.40	8.82
11	0.8761	3.0239	x					x		7257.12	7257.12	538.89	598.01	0.64	-4.84
45	0.8592	3.2985	x	x	x	x				7328.75	7328.75	792.00	1188.00	-45.09	-1027.81
24	0.8620	3.3094	x	x		x				7329.58	7329.58	792.00	1188.00	0.28	3.46
7	1.1676	3.4462	x	x						7242.01	7242.01	792.00	1188.00	0.23	23.12
23	1.1644	3.4793	x	x						7237.06	7237.06	792.00	1188.00	-37.06	-1354.27
43	2.1347	3.7936	x	x		x			x	7051.89	7051.89	792.00	1188.00	0.20	18.15
57	2.1269	3.8168	x	x	x	x			x	7045.01	7045.01	792.00	1188.00	-125.51	3941.17
25	1.0112	3.8661	x	x			x			7253.83	7253.83	792.00	1188.00	0.18	27.41
46	1.0068	3.8682	x	x	x		x			7248.47	7248.47	792.00	1188.00	-40.84	-1155.70
30	0.9911	3.9381	x		x		x			7222.78	7222.78	661.67	936.02	2.87	-1435.79
49	1.0091	4.0446	x		x		x		x	7218.57	7218.57	651.69	972.70	31.10	-2542.62
28	1.0112	4.0500	x				x		x	7221.37	7221.37	659.94	972.47	0.66	-24.00
26	1.0478	4.0781	x		x				x	7204.86	7204.86	756.24	799.98	27.70	-1780.46
8	1.0521	4.0783	x						x	7209.35	7209.35	754.54	799.50	0.48	-3.49
22	2.7471	4.1430	x	x					x	6949.42	6949.42	792.00	1188.00	-0.29	80.57
42	2.7369	4.1494	x	x	x				x	6933.06	6933.06	792.00	1188.00	-107.11	2906.50
no dist.	3.0051	4.1983	x	x	x					6942.36	6942.36	792.00	1188.00	0.00	0.00
63	1.0218	4.2712	x	x	x	x	x		x	7247.29	7247.29	792.00	1188.00	-36.65	-1442.99
59	1.0268	4.2736	x	x		x	x		x	7248.58	7248.58	792.00	1188.00	0.20	14.88
58	1.0553	4.3036	x	x	x		x		x	7224.76	7224.76	792.00	1188.00	-13.25	-2769.73
44	1.0611	4.3082	x	x			x		x	7230.73	7230.73	792.00	1188.00	0.12	35.79
33	2.0762	5.9226		x		x			x	6972.71	7051.08	792.00	1188.00	-0.01	47.90
52	2.0676	6.0080		x	x	x			x	6967.17	7048.94	792.00	1188.00	-51.12	-269.21
12	2.5105	8.3951		x					x	6803.25	6963.28	792.00	1188.00	-0.63	135.61
32	2.4939	8.4361		x	x				x	6790.49	6952.07	792.00	1188.00	-43.16	-491.12

Table B.3 – Full results of distortion test case 3: camera inclined, medium water.
Rows were sorted by values in column 3 (extrinsic reprojection error $e_{\text{extr.}}$).

Combination #	p_1	p_2	k_3	k_4	k_5	k_6	s_1	s_2	s_3	s_4	τ_x	τ_y
29	0.13	0.00	725.62	-129.29	5352.00	-667.08	0.02	-0.26	-0.14	0.52		
10	0.12	0.00	-28.03	0.00	0.00	0.00	0.01	-0.26	-0.13	0.47		
6	-0.01	0.00	930.89	0.00	0.00	0.00	0.00	0.00	0.00	0.00	0.14	0.03
51	0.05	0.08	1102.74	209.71	206.28	-1096.28	-0.09	0.39	-0.05	-0.45	0.02	0.18
20	-0.01	0.00	542.02	289.37	84.31	-542.06	0.00	0.00	0.00	0.00	0.14	0.03
56	0.00	0.00	880.86	397.12	372.64	-882.85	-0.01	0.37	0.00	-0.47	0.12	0.04
31	0.05	0.08	1295.98	0.00	0.00	0.00	-0.09	0.37	-0.06	-0.46	0.02	0.18
40	0.00	0.00	764.58	0.00	0.00	0.00	-0.01	0.34	-0.01	-0.42	0.12	0.04
41	0.03	0.02	814.07	-0.49	11.35	-816.47	-0.01	-0.05	-0.02	-1.11	0.08	0.03
21	0.03	0.02	1633.35	0.00	0.00	0.00	-0.01	-0.05	-0.02	-1.11	0.08	0.03
55	0.00	0.01	726.63	-0.35	14.58	-728.37	0.00	-0.10	0.00	-1.20	0.14	0.01
37	0.00	0.01	1456.14	0.00	0.00	0.00	0.00	-0.10	0.00	-1.20	0.14	0.01
19	0.07	0.01	28.58	-54.59	-519.93	-56.00	0.00	-0.25	-0.07	-1.15		
60	0.06	0.09	596.81	-50.26	-137.90	-586.60	-0.09	0.00	-0.06	-1.21	0.01	0.18
47	0.06	0.09	1532.28	0.00	0.00	0.00	-0.09	0.00	-0.06	-1.20	0.01	0.18
5	0.10	0.01	-1085.66	0.00	0.00	0.00	0.00	-0.37	-0.10	0.18		
15	-0.01	0.01	833.67	0.00	0.00	0.00	0.00	0.00	0.00	0.00	0.16	0.01
36	-0.01	0.01	600.40	-59.48	35.13	-601.19	0.00	0.00	0.00	0.00	0.16	0.01
50	0.00	0.00	1.19	0.00	0.00	0.00	0.00	-0.07	-0.08	0.02	-0.13	0.18
18	0.00	0.00	971.27	0.00	0.00	0.00	0.00	0.00	0.00	0.00	0.14	0.03
39	0.00	0.00	1100.54	177.53	236.33	-1103.92	0.00	0.00	0.00	0.00	0.13	0.03
35	0.07	0.00	554.23	-51.92	-116.25	-583.32	0.00	-0.06	-0.07	-1.18		
14	0.07	0.00	1383.64	0.00	0.00	0.00	0.00	-0.06	-0.07	-1.17		
54	0.00	0.00	1434.35	0.00	0.00	0.00	0.01	-0.04	0.00	-1.20	0.14	0.00
53	0.00	0.00	628.11	-57.86	97.20	-629.00	0.00	0.00	0.00	0.00	0.13	-0.01
34	0.00	0.00	695.01	0.00	0.00	0.00	0.00	0.00	0.00	0.00	0.13	-0.01
3	0.00	0.00	-17.49									
16	0.00	0.00	8.18	-0.16	43.60	28.55	0.00	0.00	0.00	0.00	0.00	0.00
38	0.00	0.00	213.12	-116.89	4771.49	-291.80	0.00	-0.02	-0.03	-0.03		
17	0.00	0.00	-109.69	0.00	0.00	0.00	0.00	-0.02	-0.03	0.08		
62	0.00	0.00	577.83	-50.04	-116.18	-578.45	0.01	-0.04	0.00	-1.21	0.14	0.00
61	0.00	0.00	86.98	-0.83	-147.10	108.58	0.00	-0.11	-0.09	0.12	-0.08	0.14
4	-0.02	0.00	-29.32	-0.19	7.52	31.42	0.00	0.00	0.00	0.00	0.00	0.00
0	-0.02	0.00	-63.19									
1	-0.07	0.00	-52.23									
9	-0.07	0.00	-161.95	-1.03	-203.43	-68.66	0.00	0.00	0.00	0.00	0.00	0.00
13	0.04	0.00	-101.68	0.69	45.32	102.82	0.00	0.00	0.00	0.00	0.00	0.00
2	0.04	0.00	-193.95									
48	0.00	0.00	2049.73	2.60	33.43	2142.80	0.00	-0.01	-0.06	0.20		
27	0.00	0.00	-64.25	0.00	0.00	0.00	0.00	-0.01	-0.06	0.20		
11	-0.04	0.00	4.13	0.00	0.00	0.00	0.00	0.00	0.00	0.00	0.03	0.05
45	0.06	0.00	19.85	-45.40	-1008.27	-20.31	0.00	-0.10	-0.04	-2.24		
24	0.06	0.00	-604.92	0.00	0.00	0.00	0.00	-0.07	-0.04	-2.30		
7	0.04	0.00	-1946.13									
23	0.04	0.00	268.62	-37.51	-1317.66	-273.88	0.00	0.00	0.00	0.00	0.00	0.00
43	0.00	0.00	-255.67	0.00	0.00	0.00	0.01	-0.11	0.11	-2.78		
57	0.00	0.00	738.08	-125.86	3983.48	-577.26	0.01	-0.12	0.11	-2.90		
25	0.01	0.01	-2141.40	0.00	0.00	0.00	0.00	0.00	0.00	0.00	0.07	0.01
46	0.01	0.01	276.60	-41.25	-1121.03	-277.69	0.00	0.00	0.00	0.00	0.07	0.01
30	-0.01	0.01	1511.44	2.15	-1402.40	1242.85	0.00	0.00	0.00	0.00	0.05	0.04
49	0.00	0.00	83.54	30.25	-2505.48	-126.89	0.00	0.00	0.00	0.00	0.05	0.02
28	0.00	0.00	276.41	0.00	0.00	0.00	0.00	0.00	0.00	0.00	0.05	0.02
26	0.00	0.00	-119.71	27.00	-1766.06	111.50	0.00	0.00	0.00	0.00	0.00	0.00
8	0.00	0.00	-120.72									
22	0.00	0.00	-1763.86									
42	0.00	0.00	252.01	-107.36	2926.05	-78.79	0.00	0.00	0.00	0.00	0.00	0.00
no dist.	0.00	0.00	0.00									
63	0.00	0.00	-173.85	-36.91	-1432.79	175.65	0.01	-0.11	0.03	-2.58	0.10	0.00
59	0.00	0.00	-1024.01	0.00	0.00	0.00	0.01	-0.07	0.03	-2.65	0.10	0.00
58	0.00	0.00	10.19	-13.65	-2744.21	-18.00	0.00	0.00	0.00	0.00	0.10	-0.01
44	0.00	0.00	-2402.62	0.00	0.00	0.00	0.00	0.00	0.00	0.00	0.10	-0.01
33	0.00	0.00	-1630.62	0.00	0.00	0.00	0.01	-0.10	0.11	-3.47		
52	0.00	0.00	-695.78	-51.19	-290.32	731.07	0.01	-0.15	0.10	-3.39		
12	0.00	0.00	-4850.43									
32	0.00	0.00	-1155.13	-43.06	-539.31	1182.82	0.00	0.00	0.00	0.00	0.00	0.00

Table B.3 – (continued).

Appendix C

Code listings

The following two Python code listings show functions used for filtering of optical action potentials, see section 5.2.3.

Listing C.1 – Python code example for Gaussian filter ignoring NaN values.

```python
import numpy as np
from scipy.ndimage import gaussian_filter

def nan_gaussian_filter(data, sigma, order=0, output=None, mode='reflect',
                        cval=0.0, truncate=4.0):
    """Filter array with gaussian kernel, ignoring pixels with not-a-number
       values.
       Inspired by: https://stackoverflow.com/questions/18697532/gaussian-
                    filtering-a-image-with-nan-in-python#36307291"""
    # Set invalid data to zero and filter in the usual way:
    v = data.copy()
    v[np.isnan(data)] = 0
    vv = gaussian_filter(v, sigma=sigma, order=order, output=output,
                         mode=mode, cval=cval, truncate=truncate)
    # Compute contribution of valid points:
    w = np.ones_like(data)
    w[np.isnan(data)] = 0
    ww = gaussian_filter(w, sigma=sigma, order=order, output=output,
                         mode=mode, cval=cval, truncate=truncate)
    # Remove contribution of invalid points:
    return vv / ww
```

Listing C.2 – Python code excerpt video filtering.

```python
import numpy as np
import scipy as sp
import scipy.ndimage
from caostk.video.bandpass import LowPass

class FilterAPs:
    def __init__(self, f_s, sigma_tex, f_iir, n_iir,l_win, h_abs, h_rel,
                 sigma_bg):
        self.samplerate = f_s
        self.spat_gauss_sigma = sigma_tex
        self.sig_lowpass_frequency = f_iir
        self.sig_lowpass_order = n_iir
        self.bg_filter_win = l_win
        self.bg_range_amplitude_abs = h_abs
        self.bg_range_amplitude_rel = h_rel
        self.bg_gauss_sigma = sigma_bg

    def filter_video(self, data):
        """Filter video"""
        n_t, n_x, n_y = data.shape
        mask_finite = np.isfinite(data[0, ...])
        mask_invalid = np.logical_not(mask_finite)
        n_finite_pixels = np.count_nonzero(mask_finite)
        if self.spat_gauss_sigma > 0.0:
            for t in range(n_t):
                data[t, ...] = nan_gaussian_filter(data[t, ...],
                                sigma=self.spat_gauss_sigma,
                                order=0, mode="constant", cval=np.nan)
                data[t, mask_invalid] = np.nan
        for x in range(data.shape[1]):
            for y in range(data.shape[2]):
                if mask_finite[x, y]:
                    data[:, x, y] = self.filter(data[:, x, y])

    def filter(self, ts):
        """Filter 1D time series of action potential (AP)."""
        # Make sure, data type is at least float64
        if ts.dtype in [np.float16, np.float32]:
            ts = ts.astype(np.float64)
        # Lowpass filter AP signal
        filter = LowPass(freql=self.sig_lowpass_frequency,
                    N=self.sig_lowpass_order, samplerate=self.samplerate)
        filter.filter_time_series(ts)
        # Estimation of resting potential in smart range:
        bg_range_top = sp.ndimage.maximum_filter1d(ts,
                        size=self.bg_filter_win, mode="mirror")
        bg_range_bot = sp.ndimage.minimum_filter1d(bg_range_top,
                        size=self.bg_filter_win, mode="mirror")
        bg_range_bot_abs = bg_range_bot - self.bg_range_amplitude_abs
        bg_range_bot_rel = (bg_range_bot - self.bg_range_amplitude_rel *
                        bg_range_top)
        bg_range_bot = np.minimum(bg_range_bot_abs, bg_range_bot_rel)
        # Mask values of signal that do lie below bg_range_bot with NaNs:
        ts_masked = np.where(ts < bg_range_bot, np.nan, ts)
        # Compute background (resting potential):
        bg = nan_gaussian_filter(ts_masked, sigma=self.bg_gauss_sigma,
                            mode="mirror")
        # Fractional Fluorescence
        return = -ts / bg
```

Appendix D

3D conduction velocity

Note: Personal copy of a memo written by Tariq Baig on April 11, 2012.
Derivation of equations for computation of conduction velocity on a curved surface.

Derivation

The basic idea is that the comoving derivative[1] of a moving isosufrace vanishes. Expressed for a scalar function F this gives

$$\frac{\mathrm{D}F}{\mathrm{D}t} = \partial_t F + \vec{\nabla} F \cdot \dot{\vec{\beta}} = 0 \ , \tag{D.1}$$

where $\dot{\vec{\beta}}$ is the velocity of the comoving frame. This can be solved either using the Moore-Penrose Pseudo Inverse or common sense to give:

$$\dot{\vec{\beta}} = -\partial_t F \frac{\vec{\nabla} F}{||\vec{\nabla} F||^2} \tag{D.2}$$

Notice that this is not the only possible solution but the one with the smallest norm or equivalently the only one completely normal to the isosurface. This is consistent with [64].

Assume now that rather than function F, whose isosurfaces are of interest, one has a function f, which is the image of F under a *time independent* mapping. Let a^i denote the coordinate functions in range of the mapping and β^j the coordinate functions in the domain of the mapping. Using the chain rule, (D.1) can still be evaluated:

$$\frac{\mathrm{D}F}{\mathrm{D}t} = \partial_t f + \frac{\partial f}{\partial a^i} \frac{\partial a^i}{\partial \beta^j} \dot{\beta}^j = 0 \tag{D.3}$$

This is invertible in the same sense as the above. Defining $T^i{}_j = \dfrac{\partial a^i}{\partial \beta^j}$ and $\overset{a}{\nabla}_i = \dfrac{\partial}{\partial a^i}$

---

[1] The derivative taken in the frame of an observer moving with the isosurface expressed in a resting frame.

we get:

$$\dot{\vec{\beta}} = -\partial_t f \frac{T^{\mathsf{T}} \overset{\vec{a}}{\nabla} f}{\overset{\vec{a}}{\nabla} f^{\mathsf{T}} T \, T^{\mathsf{T}} \overset{\vec{a}}{\nabla} f} \tag{D.4}$$

Taking f to be the 2D texture images of action potential, and the mapping to be the texture map (as defined in the texture coordinates of the mesh), (D.4) yields a method to compute the 3D CV using knowledge of the images and texture map alone.

Comparison

[45] develops an expression that can be regrouped to look similar to (D.4) using

$$M = \left[\frac{\partial \vec{\beta}}{\partial \vec{a}}^{\mathsf{T}} \frac{\partial \vec{\beta}}{\partial \vec{a}} \right]^{-1}$$

$$\dot{\vec{\beta}} = -\partial_t f \frac{\partial \vec{\beta}}{\partial \vec{a}} \frac{M \overset{\vec{a}}{\nabla} f}{\overset{\vec{a}}{\nabla} f M \overset{\vec{a}}{\nabla} f} \tag{D.5}$$

The equality of these equations would be straight forward if $\frac{\partial \vec{\beta}}{\partial \vec{a}}$ and $\frac{\partial \vec{a}}{\partial \vec{\beta}}$ would be square matrices with full rank, using the rules for transposition and matrix inversion. As this is not the case some careful inspection is necessary. As $\frac{\partial \vec{\beta}}{\partial \vec{a}}$ is a 3×2 matrix of rank 2 it has a left inverse, which is by chain rule $\frac{\partial \vec{a}}{\partial \vec{\beta}}$ since $\frac{\partial \vec{a}}{\partial \vec{\beta}} \frac{\partial \vec{\beta}}{\partial \vec{a}} = \mathrm{Id}$. By the same reasoning, $\frac{\partial \vec{a}}{\partial \vec{\beta}}$ has a right inverse. Alternatively the inverse element for a matrix A with left inverse can be computed algebraically by

$$\underbrace{(A^{\mathsf{T}} A)^{-1} A^{\mathsf{T}}}_{A_{\text{left}}^{-1}} A = \mathrm{Id}.$$

Hence $\frac{\partial \vec{\beta}}{\partial \vec{a}} M$ can be identified with T^{T} and the equality of the expressions follows.

Advantages

Expression (D.4) and its derivation is straight forward while the derivation of (D.5) is slightly convoluted.[2] Furthermore one obtains a mapping from 3d to 2d in the process of texturing a reconstructed model, using (D.4) this mapping has not to be inverted which should be of advantage in future real time applications :-). Last but not least the partial derivatives of the coordinate functions must be computed on an unstructured grid, doing this usually makes some degree of interpolations necessary. This interpolation in turn depends on the connections between the nodes of the grid which are hard to map from 3d to 2d.

---

[2] The authors do not cast it in matrix form.

Appendix E

Supplementary material

Supplementary files in digital form will be available after publication at:
`https://ediss.uni-goettingen.de/`

- Filename: `schroeder-schetelig_phdthesis_suppl_movie_001.mp4`
  This movie shows 3D panoramic optical mapping of arrhythmic activity of a
  pig heart (compare section 5.5).

List of Figures

List of Tables

Glossary

3D-EMOM 3D electromechanical optical mapping 83, 92–95, 108

AP action potential 1, 3, 4, 59, 63, 94

APD action potential duration 66

AV atrioventricular 2

BMP Research Group Biomedical Physics 4, 16, 43, 118

CG computer graphics 49, 50

CS coordinate system 97, 99, 110

CT computed tomography 6, 75, 76, 78, 104, 114, 125, 128

CV conduction velocity 4, 57, 66, 68–70

DI diastolic interval 66, 91, 93

di-4-ANEPPS Fast-response potentiometric fluorescent dye, used for optical mapping of action potentials. 13, 14, 84

diastole Phase of the cardiac cycle, during which the heart muscle relaxes. 2, 4, 59, 91, 92

DICOM Digital Imaging and Communications in Medicine 101, 103

DLP Digital Light Processing 82

DOF degrees of freedom 22, 103

DSLR digital single-lens reflex 17, 36, 84

DZHK Deutsches Zentrum für Herz-Kreislauf-Forschung e.V. 114

ECG electrocardiogram 6–8, 16, 65, 113, 118, 119, 126, 128

ECGI non-invasive electrocardiographic imaging 48, 114, 116–119, 126

EMCCD electron-multiplying charge-coupled device 14, 17, 58, 83

EPS EP Solutions SA, Yverdon-les-Bains, Switzerland 114, 115, 118

ex vivo (Latin: "out of the living") means that some biological material (cell, tissue, organ) is taken out of the organism for external examination in an environment with minimal alternation of natural conditions. 5, 7, 11, 16, 119, 123, 127

FOV field of view 97, 98, 104, 106, 126

GIMP GNU Image Manipulation Program 45–47

GPU Graphics Processing Unit 51, 53

HSL hue, saturation, lightness 46

in situ (Latin: "on site", "in position"), means that a measurement is taken where the phenomenon happens. 75, 126

in vitro (Latin: "in the glass") means that isolated biological materials are studied in a controlled artificial environment, e.g. cells in a culture medium. However, results obtained from such experiments may not be representative for or translatable to effects in the natural environment. 11

in vivo (Latin: "in the living") means that effects of various biological entities are tested on whole, living organisms. 5, 11, 114

intramural (Latin: "within the wall"), within the wall substance of an organ. 6, 75, 97, 110, 119, 128

KDE kernel density estimation 63

LA left atrium 2

LEAP low energy anti-fibrillation pacing 4, 18, 59, 65, 70

LV left ventricle 2, 108

ML machine learning 46, 47

MPIDS Max Planck Institute for Dynamics and Self-Organization 16, 115

MRI magnetic resonance imaging 114

NaN Not a Number 58, 59, 63

NrSfM Non-rigid Structure from Motion 127

OAP optical action potential 65–67

OpenCV Open Computer Vision 28, 29, 34, 35, 37–39, 153

OpenGL Open Graphics Library 28, 34–36, 51, 64, 131

PS phase singularity 67, 68

RA right atrium 2

RGB red, green, blue 44, 46

RGBA red, green, blue, alpha 51

RV right ventricle 2

SA sinoatrial 2

sCMOS scientific complementary metal-oxide-semiconductor 14

SfC Shape from Contour 40, 43, 45, 46, 53–56, 125

SfM Structure from Motion 40, 41, 43, 53–56, 125, 127

SIOX Simple Interactive Object Extraction 45

SNR signal-to-noise ratio 57–60, 63, 71, 119

SR sinus rhythm 2, 83, 84, 86, 91, 92

systole Phase of the cardiac cycle, during which the heart muscle contracts. 2, 91

TTL transistor-transistor logic 18, 101, 116

US ultrasound 97, 98, 100, 101, 106, 108–111, 126, 128

VF ventricular fibrillation 2, 68–70, 94

VT ventricular tachycardia 67, 70

YCbCr luma, blue-yellow chroma, red-green chroma 45